FRENCH LITERATURE SERIES

Le papier sur lequel le présent ouvrage est imprimé remplit les prescriptions de 'ISO 9706: 1994, Information et documentation - Papier pour documents - Prescriptions pour la permanence'.

The paper on which this book is printed meets the requirements of 'ISO 9706: 1994, Information and documentation - Paper for documents - Requirements for permanence'.

An electronic version of this volume is included in print subscriptions. See www.rodopi.nl for details and conditions.

ISBN: 978-90-420-3613-0
E-Book ISBN: 978-94-012-0884-0

Printed in The Netherlands

FRENCH LITERATURE SERIES

Since 1974 the French Literature Series has been published in conjunction with the annual French Literature Conference, sponsored by the Department of Languages, Literatures, and Cultures of the University of South Carolina, Columbia, South Carolina, USA. In addition to the scholarly papers selected for publication by the Editorial Board, it also accepts notes on the conference topic.

The conference, which is scheduled for the end of March or beginning of April each year, focuses on a pre-announced topic. The deadline for submitting conference papers is November 1; for scholarly notes, the following May 1. Submissions should be prepared according to the MLA Handbook and should not exceed fifteen pages (25 lines per page, double-spacing, with ample margins). Reading time at the Conference is limited to twenty minutes. Scholarly notes should not exceed eight pages. Authors should submit two copies of their contribution, accompanied by return postage if they wish their paper to be returned.

The essays appearing in the *French Literature Series* are drawn primarily from the Conference papers. Authors are informed of the inclusion of their papers in the volume when their papers are accepted for the Conference. Exceptionally, *FLS* does publish outstanding contributions from authors not participating in the Conference. To be considered for inclusion in the volume, such essays should not exceed twenty typed pages. A style sheet is available upon request or online at http://artsandsciences.sc.edu/dllc/FREN/LitSeries

All communications concerning the Conference should be addressed to the Conference Director, and those concerning the *French Literature Series* to the Editor, Department of Languages, Literatures, and Cultures, University of South Carolina, Columbia, SC, 29208, USA.

The *French Literature Series* is published by Editions Rodopi. Communications concerning standing orders or purchase of individual volumes or back volumes should be addressed to:

Editions Rodopi B.V.
Tijnmuiden 7
1046 AK Amsterdam-Holland
info@rodopi.nl
Tel.: 31 (0) 20 611 48 21
Fax: 31 (0) 20 447 29 79
Internet: <http://www.rodopi.nl>

USA/Canada:
248 East 44th Street
2nd floor
New York, NY 10017
USA
Tel: 1-800-225-3998
Fax: 1-800-853-3881

Future Volumes

Girls in/and French and Francophone Literature and Film (*FLS* Vol. XL, 2013)

(French Literature Series, Volume XXXIX, 2012)

THE ENVIRONMENT

in French and Francophone Literature and Film

Edited by
Jeff Persels

Amsterdam – New York, NY 2012

From the Editor

This volume of *FLS* originated with the peer-reviewed submissions selected for our thirty-ninth annual French Literature Conference (FLC) and is dedicated to James Day, long-time *FLS* editor, frequent and versatile FLC organizer, respected critic of French and Francophone literature, inspired and inspiring teacher, steadfast and exacting champion of *le bon usage*, but, above all, valued friend and colleague.

Acknowledgement goes also to the editorial board, which determined final rankings after providing at least two blind evaluations of each submission. In cases where special expertise was required, our international advisory board stood ready to provide counsel. Both the annual conference and *FLS* are indebted to the Department of Languages, Literatures and Cultures and to the College of Arts and Sciences of the University of South Carolina for their generous support.

Jeff Persels

Contents

Jeff Persels

University of South Carolina

The Environment in French and Francophone Literature and Film: An Introduction

The theme for the thirty-ninth of our annual French Literature Conferences was originally inspired by what seemed to me to be the French-speaking world's delayed engagement with ecocriticism as well as by what also seemed to me to be ecocritics' initial lack of interest in reaching back and beyond the discipline's roots in nineteenth-century English and American nature writing. The vetted contributions to that conference, collected in this volume, demonstrate, on the contrary, a thoughtful and provocative commitment to using the tools of ecocriticism both to broaden and deepen our understanding of the representation of the environment in French in works spanning a variety of genres and media — novels, short stories, documentary films, philosophical treatises, political cartoons — extending in time from the early modern to the post-colonial and in space from France to Cameroon to the entire planet.

They respond, moreover, and in creative ways, to my own hopes for ecocriticism, which I, myself, came to somewhat tardily when asked by a colleague to think about its potential for my own field of inquiry, early modern France. As it happens, this coincided with a moment of heightened anxiety about the future not merely of the planet, but, more to the point for a journal like *FLS*, for the humanities. (I write sardonically, of course, but some days, when assessing the view from an American state university office, I am not so certainof that.) Could this relatively new, environmentally conscious approach, ostensibly rooted in a specific socio-historical moment, "open up" French studies, make them "accessible and relevant, to the larger society," as German comparatist Ottmar Ette, excerpted in October 2010's apocalyptic issue of *PMLA*, advocated should be *the* survival strategy for literary studies in general? True, Ette references specifically neither French nor eco-criticism, but he does call, in particular, for attention to a greater *Überlebenswissen,* which critic Vera M. Kutzinski renders as "knowledge for survival." "For the humanities to survive in our present and future societies," Ette amplifies and Kutzinski translates, both channeling Nietzsche, "it is vital that they conceive of themselves as *sciences for living* [*Lebenswissenschaften*]" (983, emphasis hers). This is in cal–culated counterpoint to the increasing hegemony of the "life sciences" in the modern university. Literary scholars are thus encouraged to take the lead by capitalizing on their discipline's critical function to develop an open concept of life and of knowledge *about* and *for* living, systematically interrogating the "uses and disadvantages" literary scholarship has for life. Such knowledge must "serve life" – that is, it must be grounded in dialogue and theory rather than in ideology.

Ecocriticism is, I suspected then, and the articles collected here demonstrate, all over that. It very much claims to be a "science for living," a systematic interrogation of the "uses and disadvantages" of the biosphere, humanity's manipulative relationships to it, and the ways those relationships are currently, and have been historically, expressed. It is a sort of hermeneutics of the "knowledge *about* and *for* living" in harmony – or discord – with a weary planet. Drawing lifeblood from such iconic "green" writers as Thoreau, Wordsworth and Whitman and dating its "big bang" to the 1962 publication of Rachel Carson's *Silent Spring*, ecocriticism essentially addresses, as I have just observed, the relationship between humankind and the environment. Quoting from the 1996 manifesto-like *Ecocriticism Reader*, whose editors Cheryll Glotfelty and Harold Fromm themselves borrow from the mission statement of ASLE, the Association for the Study of Literature and the Environment, the ecocritical goal is "to promote the exchange of ideas and information pertaining to literature that considers the relationship between human beings and the natural world" and to encourage "new nature writing, traditional and innovative scholarly approaches to environmental literature, and interdisciplinary environmental research" (xviii). ASLE maintains an attractive and well-stocked website (www.asle.org), according to which it is "the country's premier membership organization in the field." Since 1993, it publishes *ISLE: Interdisciplinary Studies in Literature and the Environment*, once annual, then biennial, now quarterly, since 2009, which gives us some proof of the burgeoning interest, at least within the academy.

Ecocriticism is a field that, despite its recent vintage, is apparently considered by its most seasoned practitioners already to be in its second generation. These two phases may be roughly distinguished as 1) critical attention to "the personal narratives of the Anglo-American nature writing tradition" (3), in the words of Karla Armbruster and Kathleen R. Wallace whose 2001 anthology could already be titled *Beyond Nature Writing: Expanding the Boundaries of Ecocriticism*; and 2) what Armbruster and Wallace were trying to promote, i.e., "exploring the role of nature in texts more concerned with human cultures, by looking at the role of culture in nature, and by attending to the nature-focused text as also a cultural-literary text" (4). Similarly, and already in 1997, Jhan Hochman had published "an introductory critique" of what he referred to as an "emerging discipline":

> ... the project of green cultural studies is the examination of nature through words, image, and model for the purpose of foregrounding potential effects representation might have had on cultural attitudes and social practices which, in turn, affect nature itself. (qtd in Coupe 187).

In practice this means, to follow Glen Love in his 2003 *Practical Ecocriticism: Literature, Biology and the Environment*, "to reexamine and reinterpret the depictions of nature in the canonical works of the past" (34). Indeed, as former ASLE president and current *ISLE* editor Scott Slovic has expansively claimed: "there is not a single literary work anywhere that utterly defies ecological interpretation" (qtd in Love 34). Moreover, as John Talmadge and Henry Harrington observed in the introduction to their 2000 anthology, *Reading under the Sign of Nature. New Essays in Ecocriticism*: "Like feminism, ecocriticism is really less a method than an attitude" (ix). On the downside, it is true – although here it does keep company with Ette's notion of *Überlebenswissen* I invoked above – it can become a (self-)righteous attitude, even going so far as to claim to redeem the humanities. Or at least to lend them new credibility by enabling them to play a redemptive role, by proposing a "natural" way of reading, in expressed contrast to what the movement often sees as the sterile endgame of preceding theoretical approaches. To be sure, what one senses in much ecocriticism is as much a yearning or solicitude for an increasingly damaged nature as a desire for a more authentic critical relationship to it. As if the *-isms* of the past half-century were diabolically interposed between us and that authenticity, what, more aggressively, Greg Garrard has called "a coherent attack on the critical orthodoxies of today" (qtd in Coupe 182). Militant ideology did play a critical role in ecocriticism's genesis, and very much continues to do so (including in some of the articles here): "If green studies," contended Coupe a dozen years ago,

> does not have an effect on [the complacent culturalism which renders other species, flora as well as and fauna, subordinate to the human capacity for signification], does not change behaviour, does not encourage resistance to planetary pollution and degradation, it cannot be called fully "ecocritical." (4)

In support he cites one of ecocriticism's "founding fathers," Lawrence Buell, whose 1995 activist definition of the discipline was the "study of the relation between literature and environment conducted in a spirit of commitment to environmental praxis" (4-5). Ten years on, Buell was still demanding, in his 2005 *Future of Environmental Criticism: Environmental Crisis and Literary Imagination*: a "proto-ecological knowledge and environmentalist commitment" (2). Leaving aside the question of whether we can, or should, expect the literary authors we study to prove they "model ecocentric values" (Buell 22), or are "environmentally literate," we might at least ask what it would it mean for *us* to model ecocentric values and demonstrate environmental literacy when interpreting them, when mining them for a (hopefully) eco-conscious and -friendly *Lebenswissenschaft*. Widely-read ecocritic Jonathan Bate (*The Song of the Earth*) reformulated the question most usefully in 2000, in a way which bridges the gap between ecocriticism and what he calls "ecopoetics," practitioners of each of which signed on for this volume of *FLS*:

> It would be quixotic to suppose that a work of literary criticism might be an appropriate place to spell out a practical program for better environmental management. That is why ecopoetics should begin not as a set of assumptions or proposals about particular

> environmental issues, but as a way of reflecting upon what it might mean to dwell with the earth. (266)

The ten essays that follow all reflect on just that. **Jonathan Krell** opens by shifting environmental criticism to France, considering the way its French terms have been set by *the* seminal polemicized philosophical exchange in "Michel Serres, Luc Ferry, and the Possibility of a Natural Contract." **Louisa Mackenzie** marks how much more inclusive that natural contract has become in ecocriticism's progress ("It's a Queer Thing: Early Modern French Ecocriticism"), establishing provocative links between it and contemporary queer theory. **Christophe Ippolito**'s "Discours mythiques et écocritiques en dialogue: l'exemple d'Alexandrie d'Égypte," narrows focus to a specific case study, that of the myth-shrouded Egyptian city of Alexandria, to consider the literal and figurative construction of urban, cultural space. **Walter Putnam** shifts to a different sort of colonial myth-making — one of the more shameful and regrettably long-lived — the exhibition of exoticized native populations of France's far-flung empire in the contrived "natural" ecosystems of universal expositions ("Please Don't Feed the Natives": Human Zoos, Colonial Desire, and Bodies on Display"). **Claire Keith** guides us to and through the disfigured *terroir* of WWI in "Pilgrims in a Toxic Land: Writing the Trenches of the French Great War," in a survey of the post-traumatic "ecological sensibility" of the literature that unprecedentedly destructive conflict spawned. Toxicity of a different sort and its role in both environmental and cultural homogenization and degradation in colonial Cameroon provide the dominant themes of **Marie Chantale Mofin Noussi**'s "Café toxique: éco-colonialisme et monoculture dans *La Terre du café* de Patrice Nganang." **Stéphanie Posthumus** returns us to contemporary France in "Writing the Land/scape: Marie Darrieussecq's *Le Pays*," via the work of a French novelist whose sophisticated narrative devices invite (and receive) an ecocritical response. The representation of ecocriticism in the graphic lampoons of one of France's most noted political cartoonists undergoes semiotic analysis in **Liliane Toss**' "L'Environnement dans le dessin de presse de Plantu: étude sémiotique." Finall, two treatments of the innovatively filmed and distributed Yann Arthus-Bertrand eco-documentary *HOME* round out the volume. **Leon Sachs** considers the film's pedagogical message and technique in the comparison with of the influential nineteenth-century republican children's tutorial *Le Tour de la France par deux enfants*. *HOME* scriptwriter and environmental activist **Isabelle Delannoy** closes with an exposition *cum* manifesto of the film's call for new forms of expression to translate climate and environmental science into accessible and effective calls for action.

Each contribution to this edition of *FLS* thus constitutes as much a *mise en pratique* of ecocritical practice, across a variety of French-speaking and -thinking spaces and media, as a constructive *mise en examen* of ecocriticism as a vital and viable *Lebenswissenschaft*.

Works Cited

Armbruster, Karla, and Kathleen R. Wallace. *Beyond Nature Writing: Expanding the Boundaries of Ecocriticism*. Charlottesville: University of Virginia Press, 2001.

Bate, Jonathan. *The Song of the Earth*. Cambridge, MA: Harvard University Press, 2000.

Coupe, Laurence. *The Green Studies Reade: From Romanticism to Ecocriticism*. New York: Routledge, 2000.

Ette, Ottmar. "Literature as Knowledge for Living, Literary Studies as Science for Living." Trans., and with an introduction by Vera M. Kutzinski. PMLA. 125.4 (October 2010): 977-93.

Glotfelty, Cheryll, and Harold Fromm, eds. *The Ecocriticism Reader. Landmarks in Literary Ecology*. Athens, GA: University of Georgia Press, 1996.

Love, Glen A. *Practical Ecocriticism. Literature, Biology and Environment.* Charlottesville, University of Virginia Press, 2003.

Talmadge, John, and Henry Harrington, eds. *Reading under the Sign of Nature. New Essays in Ecocriticism*. Salt Lake City: University of Utah Press, 2000.

Jonathan Krell

University of Georgia

Michel Serres, Luc Ferry, and the Possibility of a Natural Contract

In *Le Contrat naturel* (1990), Michel Serres calls for humans to cease their parasitic bond to nature, and adopt a more symbiotic relationship, respecting the "objects" of the world as much as its human "subjects." Serres embraces an ecocentric vision of the world, rejecting the anthropocentric and Cartesian tradition that regards humankind as "maîtres et possesseurs" of nature. Luc Ferry maintains that *Le Nouvel Ordre écologique* (1992) must be human-centered. He ridicules Serres and those he considers "fundamentalist" ecologists, and suggests a "democratic ecology," closer to the humanism of the Classical and Enlightenment periods. Ferry's critique is not without merit, but it does not match the more nuanced arguments of Serres, Peter Singer, and Christopher Stone, all of whom envision some version of a natural contract.

> Aimer nos deux pères, naturel et humain, le sol et le prochain; aimer l'humanité, notre mère humaine, et notre naturelle mère, la Terre. (Serres, *Le Contrat naturel* 83)

> Nos conduites actuelles, notre sensibilité même, si nouvelle par rapport à la nouvelle fragilité des choses, vécues jadis comme dures face à notre impitoyable sensibilité, supposent bien que la nature devient, peu à peu, à mesure qu'elle s'affaiblit devant notre puissance, un véritable sujet de droit. (Serres, "Pour un retour au 'Contrat naturel'")

When he published his *Second Discours*, the *Discours sur l'origine et les fondements de l'inégalité parmi les hommes* (1755), Rousseau seemingly righted many wrongs of that philosophical enemy number one of many ecocritics: Descartes. Rousseau's nostalgic depiction of the state of nature, his esteem for animals, and his identification of society and laws as the foundation of inequality among human beings clearly separate him from Descartes, who, in his *Seconde Méditation* (1641), declares that humans are superior to the rest of nature, since we possess an *animus* (mind; later named *cogito*) that gives us the ability to think, whereas other living beings have at most an *anima* (soul), which might permit them to move and to feel, but never to think.

The Descartes-Rousseau dichotomy is important for ecocritics because it brings into focus two fundamental ways of viewing humans' place in nature: the classical anthropocentric model and the romantic ecocentric (biocentric) model. Michel Serres made an important contribution to the discussion with *Le Contrat naturel* (1990). This paper will examine Serres's concept of the natural contract and the scathing critique of it written by Luc Ferry in his *Le Nouvel Ordre écologique. L'Arbre, l'Animal et l'Homme* (1992). An analysis of these and other texts will lead us to ponder — if not answer — this question: could Serres's natural contract ever come to fruition, or will it remain a wistful metaphor from the pen of France's most poetic philosopher?

In *Le Contrat naturel*, Michel Serres begins with an analysis of one of Goya's black paintings, *Two Men Fighting with Clubs* (1819-23). The men are knee-deep in mud, and the viewer can imagine that as they beat each other, their efforts will only force them deeper into the quagmire as they fight to the death. Serres advises that if placing a bet on this fight, we would do well to put our money on the mud, rather than either of the two combatants. Serres interprets this painting as an illustration of the current standing of our relationship to nature. Long a victim of human aggression,[1] nature will soon be unable to endure any more, and will swallow us up, just as the mud inevitably will engulf Goya's fighters. We are about to pay for centuries of abusing and ignoring nature. "Brûlante," writes Serres, "l'histoire reste aveugle à la nature" (*Contrat*

[1] Serres differentiates between "guerre subjective" (war between people or nations, neatly defined as "violence plus quelque contrat") and "violence objective" (violence against the world) (*Contrat* 31-32).

22). For what is history but an endless recurrence of violence? As Michel Tournier writes in a brief essay entitled "L'histoire et la géographie," "Dieu dessine les contours de la géographie, mais c'est le Diable qui écrit l'histoire en lettres de sang" (89). For Tournier, geography suggests gentleness, and history evokes hate. But Serres warns that gentle space and hateful time are converging in the environmental crisis: "L'histoire globale entre dans la nature; la nature globale entre dans l'histoire: voilà de l'inédit en philosophie" (*Contrat* 18).

We find ourselves in an environmental predicament in which nature — were it capable of acting like a human — would seem to be taking its revenge. Blizzards, hurricanes, and tornadoes are more severe than ever, both droughts and floods appear to be on the rise, poison ivy and bed–bugs are stronger and angrier than ever. For all this we can thank Des–cartes, among others. Serres summarizes the Cartesian attitude towards nature in two words, mastery and possession:

> Maîtrise et possession, voilà le maître mot lancé par Descartes, à l'aurore de l'âge scientifique et technique, quand notre raison occidentale partit à la conquête de l'univers. Nous le dominons et nous l'approprions. (*Contrat* 58)

Mastery and possession have also led to pollution, as Serres explains in a more recent essay, *Le Mal propre. Polluer pour s'approprier?*. In this entertaining, albeit serious book, Serres plays on the etymological similarity between *propriété* and *propreté*, both derived from *propre*, and thus linking the concepts of cleanliness and ownership. We know how people pollute, but do we know why? Of course pollution has economic causes, but it is also part of our animal — and human — nature. Polluting something claims it as ours: if I spit in a bowl of soup, writes Serres, it's mine; if a tiger, a lion, or a dog urinates around the perimeter of an area, it has marked that territory as its own. Thus Descartes's call for thinking humans to master and possess nature ironically assimilates us to animals acting on instinct: "[Q]ui l'eût cru? [...] Descartes, le malheureux, entérinait nos usages bestiaux" (*Mal propre* 86).

It is high time, claims Serres, that a natural contract supplement Rousseau's famous social contract. Serres laments that Rousseau abandons all consideration of nature as soon as his interest passes from humans in the state of nature, in the *Second Discours*, to humans under the laws of civil society, in *Du contrat social*, written seven years later:

> Nos ancêtres les Gaulois eussent préféré, comme moi, la géographie, si sereine, à l'histoire, chaotique, et Montesquieu à Rousseau. Il faut prendre ce dernier à la lettre: passé le contrat, il n'y a plus de nature que pour le rêveur solitaire; la société l'a oubliée. Les météores s'évaporent dans les philosophies politiques, tout aussi acosmistes que les sciences sociales, après quelques premiers moments, évoqués ou pensés justement comme originaires, pour mieux éliminer le monde. (*Contrat* 117-18)

Du contrat social (1762) and *La Déclaration des droits de l'homme et du citoyen* (1789) are derived from natural law, human reason and human nature, but they are blind to non-human nature, declares Serres. These two documents, fruits of the Enlightenment and the French Revolution,

> se fermai[en]t sur soi, laissant hors jeu le monde, collection énorme de choses réduites au statut d'objets passifs de l'appropriation. Raison humaine majeure, nature extérieure mineure. Le sujet de la connaissance et de l'action jouit de tous les droits et ses objets d'aucun. Ils n'ont encore accédé à aucune dignité juridique. (*Contrat* 64)

Legal dignity, then, is what Serres aspires to confer upon nature. Human beings have been living off nature like parasites, slowly destroying their host. A natural contract would replace this parasitic relationship with a symbiotic one, essential for the survival of both humans and the world, and ending the "objective war" against nature:

> Retour donc à la nature! Cela signifie: au contrat exclusivement social ajouter la passation d'un contrat naturel de symbiose et de réciprocité où notre rapport aux choses laisserait maîtrise et possession pour l'écoute admirative, la réciprocité, la contemplation et le respect [...]. Contrat d'armistice dans la guerre objective, contrat de symbiose: le symbiote admet le droit de l'hôte, alors que le parasite — notre statut actuel — condamne à mort celui qu'il pille et qu'il habite sans prendre conscience qu'à terme il se condamne lui-même à disparaître. (67)

It is evident that Michel Serres's world view is not anthropocentric. He avoids the word "environment" in this book (*Contrat* 60), because it implies that humans are superior, at the center of creation, surrounded (*environnés*) by an inferior world. For him, humans — who could not exist without the earth — are peripheral, and the earth — which could get along just fine without humans — is central. For this reason Luc Ferry, one of Serres's strongest critics, brands him as a "deep ecologist."[2]

Ferry's *Le Nouvel Ordre écologique* is essentially a denunciation of deep ecology, ecofeminism, and any ecological movement that Ferry considers "radical" or "fundamentalist," two adjectives that recur over and over in the essay. Ferry's main critique is the anti-humanist spirit of deep ecology, which values non-human life as much as human life. This attitude, as we have seen in Serres's analysis of Rousseau's *Du contrat social*, runs counter to Enlightenment thought. Ferry would seem to favor the human-centered "shallow" environmentalism option — which is concerned only about environmental problems that pose a threat to human welfare — over deep ecology, which he considers a new and dangerous fundamentalist religion:

> [L]a première peut conserver sans dommage l'héritage de l'humanisme moderne (c'est en vertu des fins de l'homme qu'íl s'agit de respecter la terre), tandis que la seconde implique sa remise en question la plus radicale. L'humanisme ne serait pas le remède à la crise du monde industriel moderne, mais, véritable péché originel. (108-09)

Chris Cuomo, in her review of the English translation, points out many inadequacies in Ferry's book. She criticizes the "sweeping generalizations and mischaracterizations of work in environmental ethics," and Ferry's apparent view that deep ecology and ecofeminism "are pure and utter nonsense" (768). His "offhanded dismissal" of eco-

[2] The Norwegian philosopher Arne Naess is considered the founder of deep ecology. In *Ecocriticism*, Greg Garrard quotes the two crucial points of Naess's deep ecology: "The well-being and flourishing of human and non-human life on Earth have value in themselves [...]. These values are independent of the usefulness of the non-human world for human purposes. The flourishing of human life and cultures is compatible with a substantially smaller human population. The flourishing of non-human life *requires* a smaller human population" (Naess; qtd in Garrard 21).

feminism is based "on work that has been widely criticized by feminists and ecological feminists and entirely ignores the more sophisticated and comprehensive work that has been published in the 1990s" (769).

Despite these serious shortcomings, Ferry does make some valid points about the potential dangers of deep ecology, most importantly, its "moralizing power" (113) that can be transformed into a utopian, totalizing "system" (23), whose pure ambitions and fervent love for the "terroir" (23) bear troubling similarities to the *Blut und Boden* ("blood and soil") creed of the Third Reich. Like Nazism, with its roots in Teutonic mythology and its millennial dream of a new world to come after the terrible apocalyptic battle that was World War II (Eliade 25), deep ecology, in the eyes of Ferry, is obsessed with the past and the future at the expense of the present:

> [L]'écologiste profond est guidé par la haine de la modernité, l'hostilité au temps présent. [...] L'idéal de l'écologie profonde serait un monde où les époques perdues et les horizons lointains auraient la préséance sur le présent." (Ferry 146)[3]

Cuomo asserts — and I agree with her — that Ferry's most interesting chapter discusses the ecological positions of Germany's Nazi party in the early 1930s. Hitler personally recommended and had passed three important environmental laws: the first protected animals (*Tierschutzgesetz*, 1933), the second imposed limits on hunting (*Reichsjagdgesetz*, 1934), and the third, a "monument of modern ecology," according to Ferry (148), concerned the protection of nature (*Reichsnaturschutz–gesetz*, 1935). Now Ferry does not equate deep ecologists and Nazis because they share a love of nature, but he does want to point out that the Nazi example demonstrates that "la zoophilie la plus sincère" does not obviate "la haine des hommes la plus acharnée qu'on ait connue dans l'histoire" (149). Both deep ecology and Nazism share affinities with the love of unspoiled nature that characterized sentimental Romanticism and the Rousseau of the *Second Discours*. This view

[3] Jean-Christophe Rufin's novel *Le Parfum d'Adam* (2007) is an ecological thriller that explores the potential threat posed by radical ecologists. Their obsession with the past is described in the plot summary on the back cover: "La recherche d'un Paradis perdu, la nostalgie d'un temps où l'homme était en harmonie avec la nature peuvent conduire au fanatisme le plus meurtrier."

contradicts Cartesian and Enlightenment thought, for which there was no nature except that perceived and ordered by human intelligence (152), as exemplified by the formal and geometrical French garden: "le jardin à la française, travaillé, taillé, dessiné, calculé, alambiqué, artificiel et forcé est finalement si l'on veut aller au fond des choses, *plus naturel* qu'une forêt sauvage" (Catherine Kinzler, in Jean-Philippe Rameau, *Splendeur et naufrage de l'esthétique du plaisir à l'âge classique*, Paris, Minerve, 1983; qtd in Ferry 153). Before Rousseau and Romanticism, "real" nature was, Ferry adds, not wild and unspoiled, but "tamed" and "humanized" (154).[4]

Ferry concedes that Descartes's "metaphysical humanism," which considers humans "maîtres et possesseurs de la nature," has led to a "colonisation de la nature sans précédent" (25). He suggests that Cartesianism should yield to a new and improved type of humanism, "non métaphysique, non tyrannique" (25), which would make possible a human-centered "democratic ecology" (190-216), preserving humanist and democratic values while resisting the authoritarian tendencies of deep ecology. For Ferry the love of nature consists largely of "passions démocratiques, partagées par l'immense majorité des individus qui souhaitent éviter une dégradation de leur qualité de vie" (212).

Serres's vision of a contract between humans and nature is impossible, insists Ferry, for three reasons. First, nature is made up of objects, unable to express values or interests, which inevitably must emanate from humans. A natural *object* could never become a *subject* with legal rights. Attempting to do so would be a return to a "pre-modern," "pre-humanist" era (14), when beetles, leeches, rats, and other destructive creatures were actually brought to trial and threatened with excommunication if they did not mend their ways.[5] Deep ecologists who believe in the intrinsic value of nature forget that "toute valorisation, y

[4] Ferry's link of Romanticism and deep ecology to German National Socialism is well-founded. Serres reminds us, however, that the Enlightenment's social contract also has a troubling affinity with totalitarianism. The "volonté générale" under which Rousseau's ideal society would be organized could lead to the suppression of individualism and a "belle totalité" (Serres, *Contrat* 110) where neighbors would spy on neighbors in order to ensure their obedience to the state.

[5] Ferry cites several such trials that took place in France between the fifteenth and seventeenth centuries.

compris celle de la nature, est le fait des hommes et que, par conséquent, toute éthique normative est en quelque façon humaniste et anthropocentriste" (196).

Second, even if *parts* of nature hold intrinsic value, nature in its entirety does not. The biosphere contains many good things, but it also contains horrific organisms: can anyone seriously believe, asks Ferry, that the plague and cholera bacteria or the HIV virus should possess a legal status on the same level as humans (207)?

Third, if a natural contract were agreed upon, who would be the signatories of the pact? How could an unthinking concept or a collection of objects — nature — sign a legal document? Ferry suggests that Serres does not "seriously" consider that a literal contract could be possible; perhaps he merely is exercising "poetic license," and "l'idée d'un droit de la nature n'est qu'une métaphore littéraire, destinée à frapper l'attention d'une opinion publique plongée dans un sommeil lé thargique" (197). Why would Serres do this, wonders Ferry, since such lyricism surely reduces the "philosophical weight" (197) of his argument.

How have ecologists countered Ferry's objections, which do not seem entirely unreasonable? Is a natural contract, recognizing nature as a legal entity, feasible? A key refutation of Ferry actually appeared twenty years before *Le Nouvel Ordre écologique*. Back in 1972, Christopher D. Stone, a law professor at the University of Southern California, published a landmark essay on environmental ethics in the *Southern California Law Review* entitled "Should Trees Have Standing? — Toward Legal Rights for Natural Objects." Writing in defense of the Sierra Club's efforts to block Walt Disney's plans for a massive ski resort in the Mineral King Valley of Sequoia National Park,[6] Stone declares that, just as it was once "unthinkable" (459) for inanimate objects such as corporations, trusts, and nation-states to possess legal rights, the time has come for the unthinkable to happen again: for the environment to have legal standing.[7] Land, water, wildlife, and trees

[6] This case went all the way to the Supreme Court, which ruled in Disney's favor. However, the outcry from environmentalists was so intense that Disney eventually abandoned the project, and the Mineral King Valley was eventually annexed to Sequoia National Park, protecting it from development.

[7] "Throughout legal history, each successive extension of rights to some new entity has been, theretofore, a bit unthinkable. We are inclined to suppose the

cannot speak for themselves, but this is no reason for them to be denied rights. Stone proposes a "guardianship approach" (462-63), similar to that in place for humans who are no longer competent to make legal decisions for themselves. In their introduction to the reprint of Stone's essay cited here, the editors of the *Encyclopedia of Environmental Ethics and Philosophy*, J. Baird Callicott and Robert Frodeman, define the guardianship approach as follows: "an appointed guardian represents the interests of natural objects themselves, not as their degradation impacts human interests, and where monetary remedies would run to the benefit of the natural objects directly" (458).

Stone is of course an object of derision in Ferry's book, which devotes a section called "Les Arbres en procès" (18-22) to Stone and the Disney-Mineral King case. I do not find Ferry convincing, however, when he equates Stone's attempt to give natural objects legal standing to the "pre-modern," comical trials of beetles and leeches. Ferry asks his reader to ponder "postmodern" ecologists' attempts, through the legal process, to prevent humans from damaging nature because it has intrinsic value:

---

rightlessness of rightless 'things' to be a decree of Nature, not a legal convention acting in support of some status quo. [...] The fact is, that each time there is a movement to confer rights onto some new 'entity,' the proposal is bound to sound odd or frightening or laughable" (Stone, "Should Trees Have Standing?" 459).

Ethics also develop in ways that might be considered "unthinkable" to some. Decades before Stone's essay, Aldo Leopold, in the opening pages of his "Land Ethic," recounts how the ethics of ancient Greece would have permitted Odysseus, upon his return from war, to hang a dozen slave-girls merely because he suspected them of "misbehavior" (201). What would be murder today was at the time merely a man disposing of his property as he saw fit. Leopold notes that since Homer's time ethics have evolved dramatically, covering relationships between individuals (the first ethics), and later between the individual and society (the second ethics). Leopold calls for a third, a land ethic, which would include nature as well as humans: "There is as yet no ethic dealing with man's relation to land and to the animals and plants which grow upon it. Land, like Odysseus' slave-girls, is still property. The land relation is still strictly economic, entailing privileges but no obligations.The extension of ethics to this third element in human environment is, if I read the evidence correctly, an evolutionary possibility and an ecological necessity" (Leopold 203).

> Ces penseurs qui se veulent, au sens propre, "postmodernes," philosophes ou juristes de l'"après-humanisme," ne communient-ils pas étrangement dans une vision *prémoderne* du monde *où les êtres de nature retrouvent leur statut de personnes juridiques?* N'est-ce pas au fond *le même sentiment d'étrangeté* qui nous saisit, pour autant que nous sommes encore des Modernes, à l'idée que des arbres ou des insectes puissent perdre ou gagner un procès? (21-22; Ferry's italics)

Ferry earlier uses the same phrase — "sentiment d'étrangeté" — to describe his thoughts on the insect trials of long ago (17). He is obviously implying that when environmentalists thwart Disney's plans to construct a huge resort in a pristine valley, or challenge an industry's right to pollute in areas that will not affect human populations, they are performing some "strange" ritual akin to excommunicating leeches and rats. The connection is tenuous at best.

Another weak point in Ferry's argument is his claim that deep ecologists are demanding the same rights for nature that human beings possess, or that all natural objects should have the same rights. Stone makes no such claims. As if predicting Ferry's objection concerning granting rights to deadly bacteria and viruses, Stone writes:

> [T]o say that the environment should have rights is not to say that it should have every right we can imagine, or even the same body of rights as human beings have. Nor is it to say that everything in the environment should have the same rights as every other thing in the environment. (460)[8]

As for Ferry's objection that nature is incapable of signing a contract with humanity, Serres himself replies in a 2007 article that no, nature

---

[8] Peter Singer, a leading philosopher behind the animal liberation movement, similarly insists on the need to nuance the granting of equal rights to animals: "The extension of the basic principle of equality from [humans to other animals] does not imply that we must treat both groups in exactly the same way, or grant exactly the same rights to both groups. Whether we should do so will depend on the nature of the members of the two groups. The basic principle of equality, I should argue, is equality of consideration; and equal consideration for different beings may lead to different treatment and different rights" (2).

would not sign a natural contract just as no humans ever signed Rousseau's social contract:

> ai-je donc été assez animiste, assez totémiste, assez fétichiste, pour penser qu'à cette place pensait une personne? Combien objectèrent au 'contrat social' de Jean-Jacques Rousseau le même argument, puisque nul, en effet, ne le signa jamais et que nul, en effet, ne peut dater ni documenter la date et les circonstances d'une cérémonie où on l'aurait signé? ("Pour un retour au 'Contrat naturel'")

No, both the social and natural contracts are "conditions," under which humans must act "as if" a contract had been signed.

At the beginning of this essay I said that would not answer the question about the possibility of a natural contract: I will not break my promise. If Serres's proposal is viable for the philosopher or the ethicist, it poses many problems for the lawyer. In an article entitled "Environmental Law," the distinguished law professor Joseph L. Sax is not totally pessimistic regarding future legislation that could better manage and protect the environment, but he foresees at least three obstacles. First, it will be difficult to change "centuries of legal and economic development in which protecting natural systems was not a societal priority" (352). The law has long protected humans' perceived need to domesticate nature, and "thus has been the intentionally efficient cause of ecological destruction" (352). Second, the unmanaged global commons — territory unclaimed by any nation-state[9] — has and will continue to be "subject to unrestricted exploitation and to unrestricted use for waste disposal" (352). Third, political divisions and boundaries "bear no relation to ecologically rational divisions such as watersheds or habitat requirements for animal populations, the boundaries that reflect what is required for natural systems to flourish" (352). Philosophers like Serres and Ferry will continue to debate the relative merits of anthropocentric and ecocentric environmental ethics; nevertheless, if a shift is

[9] Protecting the global commons has been a particular concern of Christopher Stone. Nations would first have to accept responsibility for polluting these areas, and then an immense amount of funding through a Global Commons Trust Fund would be required. See *The Gnat is Older than Man* and "Defending the Global Commons" in Sands, *Greening International Law*.

ever made from the first to the second, legal recognition of such concepts as a natural contract, the intrinsic value of nature, and the perception of nature as a living organism will require a "vast leap" (352) by environmental legislators.[10]

Works Cited

Callicott, J. Baird, and Robert Frodeman, eds. *Encyclopedia of Environmental Ethics and Philosophy*. 2 vols. Detroit: MacMillan Reference USA, 2009.

Cuomo, Chris. Rev. of Luc Ferry, *The New Ecological Order*. Trans. Carol Volk. *Isis* 87.4 (1996), 768-69.

Descartes, René. *Méditations métaphysiques, t.1-3*. Paris: Gallimard, 2006.

Eliade, Mircea. *Mythes, rêves et mystères*. Paris: Gallimard, 1957.

Ferry, Luc. *Le Nouvel Ordre écologique. L'Arbre, l'Animal et l'Homme*. Paris: LGF, 2002.

Garrard, Greg. *Ecocriticism*. London: Routledge, 2004.

Harris, Paul A. "The Itinerant Theorist: Nature and Knowledge/Ecology and Topology in Michel Serres." *An Ecology of Knowledge: Michel Serres*. Ed. Sydney Lévy. Spec. issue of *SubStance* 26.2 (83) (1997): 37-58.

Leopold, Aldo. "The Land Ethic." *A Sand County Almanac*. 1949. Oxford: Oxford University Pres, 1987. 201-26.

Lovelock, James. *Gaia: A New Look at Life on Earth*. 1979. Oxford, Oxford University Press, 1987.

Paulson, William. "Literature, Knowledge, and Cultural Ecology." *Épistémocritique*. Ed. Noëlle Batt and Michel Pierssens. Spec. issue of *SubStance* 22.2,3 (71/72) (1993): 27-37.

______. *The Noise of Culture: Literary Texts in a World of Information*. Ithaca: Cornell University Press, 1988.

[10] The "Gaia Hypothesis," developed by James Lovelock in *Gaia: A New Look at Life on Earth*, considers the entire earth to be one living super-organism. Like Serres's natural contract, Lovelock's hypothesis has been ridiculed by some as a "charming metaphor" (Garrard 174).

Rousseau, Jean-Jacques. *Œuvres politiques*. Paris: Bordas, 1989.

Rufin, Jean-Christophe. *Le Parfum d'Adam*. Paris: Flammarion, 2007.

Sax, Joseph L. "Environmental Law." Callicott and Frodeman. Vol. 1. 348-54.

Serres, Michel. *Le Contrat naturel*. Paris: Flammarion, 1999.

______. *Le Mal propre. Polluer pour s'approprier?* Paris: Éditions le Pommier, 2008.

______. "Pour un retour au 'Contrat naturel'." *Le Figaro*. 17-18 Nov. 2007. <http://www.lefigaro.fr/debats/2007/11/16/01002007116ARTFIG00321-environnement-pour-un-retour-au-contrat-naturel.php>

Singer, Peter. "All Animals Are Equal." *Animal Rights and Human Obligations*. Ed. Tom Regan and Peter Singer. Upper Saddle River, NJ: Prentice Hall, 1989. 148-62. <http://www.animal-rights-library.com/texts-m/singer02.htm>

Stone, Christopher D. *The Gnat is Older than Man: Global Environment and Human Agenda*. Princeton: Princeton University Press, 1993.

______. "Defending the Global Commons." *Greening International Law*. Ed. Philippe Sands. NY: The New Press, 1994. 34-49.

______. "Should Trees Have Standing? — Toward Legal Rights for Natural Objects." *Southern California Law Review* 45 (1972): 450-57. Rpt. in Callicott and Frodeman. Vol. 2. 458-84.

Tournier, Michel. "L'Histoire et la Géographie." *Le Miroir des idées*. Paris: Mercure de France, 1996. 85-89.

Louisa Mackenzie
University of Washington, Seattle

It's a Queer Thing: Early Modern French Ecocriticism

Ecocriticism, or the study of nature-culture intersections, has expanded its critical and geographic scope well beyond the study of modern American and English nature writing with which it originated. Ecocritics increasingly pay attention to issues of race, class, ethnicity, gender, sexuality, and national identity in their interrogations of cultural inflections of nature, and are also starting to consider pre-modern periods. This article builds on three of these many developments: early modern ecocriticism, a specifically French *écopensée*, and queer ecology, suggesting a multi-dimensional, articulated approach that can help us continue to question the premises of the nature-culture divide.

Although the term "ecocriticism" can be traced back to William Rueckert in 1978, one of the now-canonical departure points for those seeking to define it is Cheryl Glotfelty's study of "the interconnections between nature and culture, specifically the cultural artifacts of language and literature" (xix).[1] Since 1996, when Glotfelty's landmark ed-

[1] Many ecocritics also include "a spirit of commitment to environmentalist praxis" in this definition (Buell 430 n.20), although see Phillips (160-72) and Morton (*Ecology*) for a critique of this notion of praxis as it relies on "realism." The word "ecocriticism" itself is not without its critics; some including Buell prefer other options such as "environmental criticism." For excellent summaries of developments in the field, see Heise ("The Hitchhiker's Guide") and Buell ("Ecocriticism"); also Garrard (*Ecocriticism*) and Buell (*Environmental Imagination*) for book-length studies. Buell's understanding of ecocriticism's

ited collection appeared, ecocriticism has been contested territory. Productive contestation is, of course, a necessary part of the evolution of any *-ism*, without which it ossifies and fades into irrelevance. Ecocriticism since the 1990s has shown itself to be a robust and adaptable critical approach, not easily dismissed as a fashionable academic moment. Perhaps the true test of a field's critical potential is whether it can evolve in ways that are not merely additive, but that change some of the very premises of the field itself. Such has been the case with ecocriticism's responses to interventions from feminisms, critical race studies, non-Anglophone or non-Western traditions, cultural geography, etc. These have not simply added extra limbs to an unchanged critical core; rather, they have profoundly complicated and nuanced the core itself. To put it differently, ecocriticism is not just about the intersection between human and non-human worlds, but is also a productive and mobile site for intersectional ideologies and theories. I will here suggest that three recent interventions in ecocriticism — from early modern studies; from a French traditional of ecological thinking which, following Stephanie Posthumus, I will call *éco-pensée*; and from queer theory — can each be brought into dialogue not only bilaterally with ecocriticism but also with each other, producing a multi-directional articulation that enriches our understanding of each. More specifically, this articulation can help us to continue to theorize hybridity in nature-culture relations. The notion of the radical separateness of non-human nature and human culture continues to haunt some ecocritical practice and environmental activism. This "ecocentric versus anthropocentric" debate — characterized as "nature-endorsing" versus "nature-skeptical" by Soper (4) — is one of the core impasses in environmental humanities in general, despite many attempts to think beyond the binary (MacCormack and Strathern, White).[2] It has led to "theory wars" between

---

history relies on a wave model; in feminist history, wave models have been challenged (e.g., Laughlin et al.). Such challenges might also be applied to the notion of ecocritical waves.

[2] There are numerous examples; in fact it now seems almost mandatory to present a study as one that challenges the binary. One might wonder whether the binary reproduces itself partly through academic insistence on overcoming it (including my own). For an anthropological approach which argues that the concepts of nature and culture co-produce each other in modernity, see Vandenberge.

scholars which tend to dead-end,[3] and also underpins a kind of ecopraxis which is arguably less effective than one based on a hybrid nature-culture. Nature-culture, or natureculture, is certainly not new: it is central, for example, to the work of Donna Haraway and Bruno Latour. What I am proposing is a way to apprehend hybridity by approaching it from the simultaneous approaches offered by early modern studies, *éco-pensée*, and queer theory, all of which intervene critically to complicate the notion of a nature/culture (biotic/social; ecocentered/anthropocentered) exclusivity — and which, in turn, can themselves be nuanced by an encounter with the ecocritical.

It bears repeating that ecocriticism is not a monolithic ideology or approach. It has on occasion been unfairly and simplistically set up as a straw man (Phillips, T. Morton).[4] There is some validity to these critiques, but they tend to focus on one kind of ecocritical practice that was more the norm in the 1990s, which was dominated according to Patrick Murphy by an "anti-theoretical, naïve, realist attitude" (165). Wed to environmental activist ideals, some ecocritics identified reading, writing, theory, books, and postmodernism in particular, as part of what was alienating us from the natural world (Kroeber) and proposed that "theorists should go outside and feel the rain on our heads every so of-

---

[3] Feder notes that "those hostile to theory tend to treat it as monolithic" (775); it should be noted that some who embrace theory can also do so in an ideologically monolithic way. Soper provides a still-relevant reminder that "both 'nature-endorsing' and 'nature-sceptical' perspectives need to be more conscious of what their respective discourses on nature may be ignoring and politically repressing" (8). The ongoing tensions between ecocriticism and theory are the object of a recent special forum in the field's flaghip journal, *Interdisciplinary Studies in Literature and Environment* (ed. Slovic). Not all early environmental critics' suspicion of theory can fairly be characterized as naïve: Buell's *Environmental Imagination* is qualified and more nuanced than given credit for by his detractors. Simon Estok challenges the idea of theory as abstract and elitist, arguing hopefully that it can help "cross the Great Divide between intellectuals and the world" ("Bridging" 207).

[4] In his 2003 polemic, Dana Phillips blasts the field for what he sees as its naïve realism, its failure to be interdisciplinary, and its attachment — despite its purported interest in the world beyond books — to old-fashioned notions of literary criticism. More recently, in his 2010 prolegomenon for a "queer ecology" (to which I will return), Timothy Morton accuses ecocriticism of being radically masculinist, heterosexist, and attached to simplistic binary systems.

ten" (Slocombe 503) instead. Enthusiastic claims were made for ecocriticism's potential to green the academy, bring about new holistic forms of knowledge (Kroeber, Branch), and forge new interdisciplinary connections. Certain canonical works of U.S. nature writing or British romanticism were the predominant objects of study (Bate 1991, Slovic 1992) and sometimes held up as embodying an ideal or lost relationship between humans and nature (Love, "Revaluing"). This branch of ecocriticism privileged notions of wilderness, unsullied environments, and an attachment to a unitary nature as absolutely beyond culture.[5] Urban and more hybrid settings did not qualify as environmental (Sanders). A certain minority of ecocritics is still very attached to the idea of an absolute separation between nature and culture; some are angered at any suggestion that nature is cultural.[6] There are reasons for this, which have to do with ecocriticism's admirable activist engagements: we want to protect the environment from the depredations of human activity. And we think that humanizing the environment — that is, recognizing the cultural filters through which we inevitably think about nature[7] — amounts to epistemological domination and thus exploitation. But the appreciation of wilderness on which this position has tended to rely, is problematic. Wilderness is itself a contested notion (Cronon). There is explicitly no place for the weak and implicitly no place for the female, the queer, the non-white, the poor, in the injunction to "[s]hut up and go outside" (Robisch 702), with all its performance of rugged masculinism[8]; no awareness that the very ability to build "trail cred" (are trails

---

[5] An exception is Wilson , who makes the case for "an idea of nature that includes human culture" (17).

[6] For an extreme example, see Robisch, who dismisses such ideas as theoretical "masturbatory apparatus," and imagines showing up at conferences to throw paint on "these nature fakers, these seated hikers." As Katie Hogan points out, "for queers, as for many racial, ethnic, gender, and religious minorities, nature is not a hiding place from ideology but often its location" (237).

[7] This is the problem of "sign pollution" discussed by Slocombe: "even ecocritical readings of literature and nature are still part of human discourse, and are therefore as susceptible to linguistic slippage and anthropocentricism as postmodernism" (502). While we can only ever discuss or represent the real through signs that are always already polluted, we do inhabit this real "bodily, even if not textually" (502).

[8] Heise describes the focus on restorative relations with environment in nature writing as "the experience of single, mostly male individuals encountering wild

not constructed?) is determined by the individual's culture, gender, bodily ability, class and privilege. Second, and this is more my focus here, it tethers discussion to a kind of binary impasse, where *culture* is devalued because associated with human activity, and *nature* is a persecuted other-out-there to be protected. And this notion of a pure unsullied nature is itself a cultural construction (whose political Eurocentric uses are explored brilliantly by Latour in *Politiques de la nature*),[9] leading to a brand of environmentalism which is unrealistic at best, radically exclusionary at worst.

To continue to accuse ecocriticism of naïveté, however, seems decidedly out of touch, ignoring the ever-increasing complexity and sophistication of the field. It is always possible to prove a stereotype if that is what one is looking for, and as Michael Cohen points out, it is too easy to dismiss ecocriticism as the "work of scholars who 'would rather be hiking'" (12). Today, ecocriticism is less white, less male, less blind to privilege, less exclusively housed in English departments, than it was in the 1990s. It has moved outwards and forwards in many ways, gained in critical self-awareness, with interventions from feminist and queer theories, critical race studies, postcolonialism, environmental and social justice, and some non-Western literary traditions.[10] Leading early ecocritics such as Buell, Slovic, Love, and Branch, are themselves significantly rethinking the parameters and premises of their own work. "Nature" is now understood by most ecocritics as a profoundly contingent concept, traversed by social, cultural and political forces. It is now

---

landscapes or homesteading agricultural ones" ("Ecocriticism" 385). Mei Mei Evans makes a similar point (181-84).

[9] Although it is important to note that Latour specifically rejects the notion of social construction (50-62).

[10] Among many others, Adamson et al., Alaimo, Gaard, Huggan and Tiffin, Murphy, Outka, Sandilands, Sturgeon, Myers, Vital, have written or edited book-length studies on intersections between environmental concerns and one or more of those mentioned. Buell (2011) provides an excellent recent overview, with copious bibliographic information in the footnotes, of important new developments. These studies are all situated, intersectional studies which avoid conflating different mechanisms of oppression which would too simply equate racism with sexism with homophobia with exploitative attitudes towards nature, etc. (as some early eco-feminists, e.g., Merchant, grafted misogyny onto the will to dominate nature).

hard to posit nature as a transcendent extra-cultural experience available to individual subjectivities unaware of their relative privilege: a recent issue of *Qui parle* on ecocriticism is presented as analyzing "dynamic conditions of interaction, entanglement, and collectivity: ecology, environments, biomes, landscapes, naturecultures, second nature, vital materialism, planetary garden" (Dodson 7). In this and other ways, ecocriticism increasingly shares modalities with anthropology and cultural geography, which interrogate the cultural nature of landscape, space and place (e.g., Cresswell in English, Roger in French).

I am about to suggest possible new articulations for ecocriticism, then, but not because it is currently impoverished; on the contrary, it is a very productive site for new kinds of conversations. Ecocriticism's ability to "adhere to a social and political program" comes from an ability to accept, not reject, "critique of the way it structures ethical issues" (Cohen 12). An increasingly hybrid, portable, humanized, and politically engaged kind of nature can only strengthen ecocriticism's potential both as a theory and as an ethic. It's a good time to think about an early modern, a French, and a queer ecology.

Despite the many ways in which it has diversified, ecocriticism is still largely focused on cultural productions in English, as Buell himself recognizes at the end of his excellent exposé of recent trends in the field (107) and from the nineteenth century or later. In what follows, I will suggest that ecocriticism can be put in concurrent dialogue with 1) early modern studies, 2) French philosophies of nature, or, as noted above, what Stephanie Posthumus has termed *éco-pensée*, and 3) queer theory. This might seem to be a grotesque hydra: an early modern/French/queer ecocriticism constructed in a purely additive spirit. But each of these fields has already, individually, intervened to make ecocriticism an even more robust and portable approach to culture. Ecocriticism can be *simultaneously* early modern, French, and queer, and that this kind of multi-dimensional, articulative theoretical model (going beyond the implicit dualism of "ecocriticism+queer theory," "ecocriticism+French philosophy," or "ecocriticism+early modern culture") can continue to powerfully complicate a binary mode of thinking about nature and culture. And in return, ecocriticism can add important new directions to early modern studies, to French thinking about nature, and to queer theory.

These three areas will be presented one by one, in a linear way, but also gesturing towards articulation, in the three-dimensional, non-linear sense. First, there is the relative lack of focus on what is not unproblematically called the "early modern" period.[11] Recently, ecocritical interest has burgeoned in English early modern history and culture (Watson, Estok, Borlik, Theis, Hiltner), but it is still an emerging, albeit very lively, field, and there is almost nothing on non-Anglophone early modernity (which reflects the origins of ecocriticism as a whole). One of the tethers of the term "early modern" is that even subtle ecocritical studies aware of the problems of positionality can tend to apprehend the period's cultural products inasmuch as they anticipate or contradict our own modernity. This notional modernity and its depredations on the environment has become something of a narratheme among ecocritics: it is still widely held that a modern scientific mentality has increasingly alienated humans from nature.[12] But what if, as Bruno Latour so cogently proposed, we have never been modern? If there were no radical separation from our past, or from nature, occasioned by modern science? Latour proposes that science is constituted of hybrid systems of knowledge, "hybrides de nature et de culture" (21), and that this hybridity is central to, while also denied by, what he calls the modern constitution.[13] To collapse the nature-culture binary, for Latour, is also to collapse the ancient-modern binary. In other words, if we have never been modern, perhaps we have always been early modern.

Early modern European writers, whether scientists or humanists — although Anthony Grafton has amply shown the ways in which our disciplinary distinctions are certainly not theirs — were Latourian thinkers. Nature was understood as always already philosophical and theological, *natura naturans* (what we might now think of as "mother nature," a creating force), or *natura naturata* (that which is created by

---

[11] See Brown for the problem of periodization in general, and Marcus for the early modern/Renaissance debate. In French literary history written in French, *Renaissance* is still more or less synonymous with the sixteenth century.

[12] However, ecocritics are starting to consider "new technologies as not necessarily oppositional to ecological frameworks" (Dodson 9).

[13] See chapter 2 of Latour's *Nous n'avons jamais été modernes* on the idea of a modern constitution, the relation between hybridity and "purification" in knowledge systems.

this force, possibly and vaguely analogous to "environment").[14] Whether writers described physical environments or a more philosophical *Natura*, they did so in a way that took for granted that "culture is inseparable from the materiality of nature" (Theis 43). According to the "Great Chain of Being" world view so enduringly described by Lovejoy, human and non-human natures coexisted as part of an intricately connected hierarchy which led all the way up to God. The ways in which this justified massive social inequalities have since been pointed out, but Todd Borlik argues that it may be time to revisit the possibilities of its "'analogical habit of mind' that can be seen as a forerunner of an ecological sensibility" (11). And while there are early modern descriptions of environmental change, and versions of the Bad Modernity myth, these cannot be understood primarily as activist protest in the modern sense. There is certainly evidence, in many early modern cultural products, for a will to conserve natural spaces. But this awareness of destruction should also, and simultaneously, be considered as part of concurrent philosophical traditions such as that of *perpetuum mobile*, and the nature threatened is not the unitary, non-human nature that it is today.[15]

Nature is, as Raymond Williams famously wrote, the "most complex word in the English language" (219): exploring and even celebrating that complexity does not make us anti-environmentalists, any more than being a Biblical exegete diminishes a believer's faith. Timothy Morton has suggested that we get rid of the term *nature* — the unitary and separate nature he identifies in most ecocriticism — in favor of the more interconnected notion of ecology: "the idea of nature is getting in the way of properly ecological forms of culture, philosophy, politics, and art" (*Ecology* 1).[16] We should not, of course, abandon Nature, but rather

[14] There are numerous studies on Renaissance notions of nature; a helpful overview is Maclean. For an excellent reading and contextualization with respect to a French sixteenth-century writer, see Blair.

[15] See Borlik for a demonstration of the centrality of lived environmental concerns to a variety of early modern English texts, and Jeanneret for the dominance of the *perpetuum mobile* paradigm. I am arguing here that both can be true, indeed they must be; early modern texts require us to suspend our process of separation between biological and philosophical.

[16] I wonder if, in his programmatic evocations of notions of interconnectedness as synonymous with ecology, Morton might not be performing some of the troping on ecological metaphors of which Phillips accuses ecocritics (42-82,

give due place to early modern *natura*, a humanized nature that resists a nature-culture divide. And we must — indeed our own historical moment will continue to demand it — remain attentive to the contexts of privilege and ideology that produce historically shifting natures. That is, in revisiting early modern *natura,* we cannot assume it to be any more true or inclusive than modern nature; we must consider whose power was supported by such notions (which would include, for example, the church).[17]

It is not just ecocriticism that is challenged by the encounter with the early modern. The reverse is true, too. Readings of early modern (in this case French) texts can be enriched by a historically appropriate ecocritical theory: we can use the latter to look beyond the densely intertextual referential universe of early modern descriptions of nature to try to understand something of how lived experience of environments and written descriptions thereof worked together. While scholars of early modern literature should be particularly hesitant about the category of realism, we might have become overly conditioned to see early modern literature as an echo chamber of mediated, erudite allusions.[18] This is particularly true in French early modern studies: the English tradition, influenced by Raymond Williams and cultural geography, has been more receptive to articulations between text and lived environments, (e.g., Fitter, Hiltner). Early modern writers did of course live in, enjoy, react to, and feel, physical environments.[19] Poets such as Pierre de

see also Heise, "Hitchhiker" 510). In fact we all might be guilty of this to some degree, whether we evoke connectedness as part of a holistic deep ecology worldview, or as part of how we might imagine queer ecology.

[17] There is interesting work attempting to revisit the category of the sacred in nature. See McGrath for a Christian natural theology, and Frank (an astrophysicist) who makes a non-denominational case that the science-religion conflict is a red herring.

[18] Hugely influential studies such as those by Cave and Greene have shown how central intertextuality and imitation are to writing practice, and how necessary for interpretation. While their deconstructive and intertextual turns were crucial in moving away from a somewhat naïve paradigm of literature "reflecting" some extra-textual truth, their hegemony has arguably tethered us to a hall of mirrors in which any reference to non-literary external realities is treated with suspicion.

[19] The obviousness of such a statement perhaps also brings home the slightly silly nature of the current stereotype of ecocritical theorists as people who just

Ronsard and Antoine de Baïf witnessed and deplored actual environmental change (deforestation and water pollution, respectively), although both write about it in densely mythological-poetic-allegorical terms, and from the perspective of aristocratic landowning humanists. Bernard Palissy argued for moderate exploitation of natural resources, while the humanists Agrippa and Agricola debated the morality of mining. Maurice Scève and many others debated the merits of country versus city living. There is currently no interpretive practice that allows us fully to understand the import of literary references to, and transformations of, lived experience of the environment. A historically sensitive ecocriticism might provide just such a practice.

As mentioned above, the move towards early modernity in ecologically-oriented criticism has to date happened mainly in English departments, which reflects the origins of the field. With my research focus being French, it has been difficult to pursue an interest in ecocriticism (in any period) without ending up in a primarily Anglophone discursive universe: there is very little in French, treating French texts, that corresponds (although Sampon-Nicolas proposes a French eco-canon that spans from Ronsard to Ponge). There is presently a significant internationalization of ecocriticism (there are chapters or affiliates of the *Association for Literature and the Environment* in India, Japan, Korea, and elsewhere, many of whose members do work on non-English topics), but most ecocritical work is still published in English, and predominantly on English texts through an Anglo-American critical lens.[20] These problems are distinct but closely related: the question of national focus is being addressed faster than the question of monolingualism which, for Heise, does not allow us to "think globally [...] especially when that language is a hegemonic one" ("Hitchhiker" 513). It is not clear whether the prevalence of English as the default language of conferences or publications can be substantially addressed (that is, in

---

need to go outside and get wet in order to understand that "ideas are no big deal" (Rogers, 12). Having ideas (or being a sixteenth-century poet) does not erase bodies, or stop them from getting wet, or make us any less awestruck by nature's beauty, power, etc. Neil Evernden has usefully suggested getting over the ivory tower metaphor (ix-x), pointing out the falseness of the assumption that academia is somehow not part of "the world" and its ecologies.

[20] Posthumus argues that the internationalization of ecocriticism is mostly about making an English canon and approach known abroad, a subtle form of Anglophone monoculturalism.

a non-tokenistic way) other than by a willingness on the part of Anglophones to listen to those speaking in and for non-English sources, and to recognize the ways in which language and national origin determine our conceptions of the natural. Playing devil's advocate, one could argue that if ecocriticism originated in English and American literature departments, and treats primarily Anglophone culture, it could continue to live happily where it is, taking into account — as it has done — the need for more sensitivity to issues of gender, race, etc. But I believe that to put ecocriticism in dialogue with different national traditions is not to play some kind of uninvited anti-Anglophone trump card, nor does it have to be about exporting a monolithic ecocriticism. In terms of the specificity of a French tradition of thinking about ecology and nature, it can create with Anglophone ecocriticism a productive dialectic, and both can become more critically and practically compelling from the articulation.

To date, English ecocritical engagement with French thought has almost exclusively focused on texts already translated into English — such as seminal works by Derrida and Foucault, which turned critical attention to the apprehension of reality through discourse — which are then instrumentalized as part of a justification, or rejection, of theorizing in general.[21] Many ecocritics note that developments in French critical theory from the 1960s to 1990s, tended to approach nature as a social construct. One citation which has become something of a meme is Roland Barthes's proposal back in 1957 "always to strip down Nature [...] to posit Nature as itself historical" (quoted and translated by Heise, "Hitchhiker" 505).[22] Although Barthes did not mean abandoning respect for ontological, biological nature, but rather accepting that any way in which we humans discuss and experience nature must be filtered through human cognition, this kind of thinking was decried by many Anglophone environmental scholars as evidencing a disregard for na-

---

[21] It must be noted that French texts considered seminal to critical theory in the Anglophone academy are not necessarily so in France. For a thoughtful study on the importing (and sometimes the deformation) of French theory in the Anglophone academic world, see Cusset.

[22] The English translation of this passage is also cited, among others, by McGrath (111). Floating citations of French theory in English are quite common among ecocritics, e.g., Oppermann quoting Vandenberge quoting Debray (769).

ture's existence beyond the human realm.[23] "Theory" became symptomatic of an alienation from, and hence indifference to, the reality of environmental degradation (Buell *Environmental*, Sagar for a more radical example). Thus, in a process that can perhaps be compared to the ongoing debates between Francophone and Anglophone feminists, French thought has been seen in the English world as problematically apolitical, while suspicion towards American pragmatism and "le politiquement correct" endures in France.[24]

In a recent attempt to put the two in dialogue, Stephanie Posthumus argues that French *éco-pensée* (her term) "answer(s) some of the questions about ecology, ethics, language, modernity and realism that ecocriticism has been asking since it first emerged" (1). She draws largely on the work of Michel Serres, arguing that his career-long search for inter- and trans-disciplinary connections provides a model for what ecocriticism could itself be doing; she also explicitly reads Serres as a humanistic ecological thinker. In this and other ways she makes a convincing case that Serres (much of whose work has not been translated) can provide a French-specific intervention in ecocriticism, rather than "simply applying North American environmentalism to literary texts from other cultures" (6). Luc Ferry, in an over-simplification of Serres that sets him up as an anti-humanist straw man (whereas in fact Serres argues throughout his *Contrat naturel* for a symbiotic, *reciprocal*, relation to nature), has provided a spirited defense of humanism against what he sees as the excesses of some environmental movements, including deep ecology (e.g., Naess) which influenced a lot of first-wave ecocritics. Ferry often overstates his case, making rather

[23] See Johns-Putra, 745-46, for a summary of this position among leading ecocritics Buell, Garrard, and Love.

[24] Recent media exchanges, provoked by "l'affaire DSK" (the arrest in May 2011 of Dominique Strauss-Kahn on rape charges), on sexism in French political culture have specifically continued the opposition between French and American feminisms, with the latter characterized as militant and politically correct by the French, and the former as internalized misogyny by Americans. The French-American feminist and Renaissance scholar Cécile Alduy has provided superb analysis of the cultural differences raised by this scandal in her various media interventions.

glib connections between deep ecology and Nazism,[25] but his main point — that humanistic environmentalism is more likely to succeed and in no way has to detract from the wellbeing of the biosphere — is well taken. Anglophone ecocriticism has no such manifesto in favor of humanism: the closest might be Castree and Braun's 2001 collection of essays *Social Nature*, but all contributors are geographers.[26]

French environmentalisms seem to be always already humanistic, and refuse for the most part to engage a nature-culture opposition (Descola). This is the main point made by Kerry Whiteside, who, as Posthumus does for Michel Serres, introduces a panoply of French environmental thinkers to an Anglophone readership arguing that they variously posit a "noncentered ecological argument" from which the anthropocentric-ecocentric debate is largely absent (3).[27] Bertrand Westphal proposes a model of "géocritique" which sees imaginary and biotic worlds, place and text, as a fundamental and ongoing process of relationality. Text, for him, is not an epiphenomenon, but exists in perpetual "incessante évolution" (186) with physical place.[28] Such models of cultural analysis can challenge ecocriticism's uncomfortable relationship to culture and representation.[29] And it is a two-way street: Anglo-

---

[25] For a thoughtful analysis of the relations between Heidegger, Nazism, and deep ecology, which acknowledges the problem without making it a reason to dismiss ecological thinking, see Garrard, "Heidegger."

[26] Evernden has argued for the need to appreciate the social creation of nature (a development of his notion of environmental phenomenology), but more in order to appreciate how we get "nature" wrong when we try to identify what needs to be saved; Evernden remains deeply attached (*pace* Robisch) to the notion of wildness.

[27] The need for further conversations between environmental theories and early modern studies is seen in an arguably trivial but revealing detail in the second sentence of Whitehead's introduction, where he presents Descartes as a "sixteenth-century Frenchman" (1).

[28] Westphal engages throughout the significant terminological debates about space versus place, *espace* versus *lieu*, too various to summarize here; for a helpful introduction, see Cresswell.

[29] Ning Wang has recently made a similar case with respect to ecocriticism in Chinese contexts (2009). Some other relevant studies in French are Blanc, Pughe, Larrère, and the volume of *Esprit Créateur* dedicated to "éco-littérature" (Desblache).

phone ecocritical activism, if reformulated in a culturally specific way (that is, not as an overtly American criticism applied to American texts), might encourage the philosophical conversations about culture and the environment in France (Ferry, Baudrillard), to be less impatient with anything that to them smacks of American nature-worship. The work of Catherine Larrère suggests as much: directly addressing the differences between French and American ecological thinking, she argues in favor of an American "sens moral" ("Éthiques" 76) which would respect plurality and not oppose nature to culture, an "éthique écocentrique qui a l'ambition d'intégrer les activités humaines dans l'environnement naturel" (84). The work of the *Esthétique environnementale* network in France (see Blanc for an overview) is interested in environmental experience "à l'intérieur de soi et non plus [...] l'objet d'une contemplation passive" (Blanc 11). It should be noted that they do not perceive Anglo-Saxon environmentalisms as rigidly ecocentric; rather, the interest for them is to learn from traditions of US civic action in order to blur the categories of human and natural (Blanc 16). Thus, the comparison I am making between French and English ecocriticisms is itself rather schematic, and would need to be refined by the kind of specific studies modeled by, e.g., Posthumus.

So what, finally, might be queer about any of this? It is too easy to claim to have queered something simply by virtue of complicating a binary, and queering can sometimes be a simply additive process of "add[ing] queers and stir[ring]" (Sandilands, "Lavender's Green" 21). But I share with Timothy Morton the intuition that "fully and properly, ecology is queer theory and queer theory is ecology" ("Queer" 281).[30] Queer ecology is not coterminous with gay ecology, that is, it's not simply outing nature by pointing out the ways in which the natural world is not always heterosexual, or critiquing the "naturalization" of certain be-

---

[30] Morton's article is productively provocative; however he gives only passing mention to Mortimer-Sandilands and to Gaard, whose work has been breaking this new ground since the 1990s. His characterizations of the rugged individualism and hypermasculinity of ecocriticism, and of ecofeminism as a direct descendent of "feminist separatism" (274), do not take into account the multiple expansions of the field in the last 20 years. See Mortimer-Sandilands and Erickson (1-47) for an overview of the stakes and the development of queer ecology to date.

havioral regimes such as the straight nuclear family.[31] Indeed, the power of "queer" is precisely in the fact that it's not just about sex (and when it is, it's not just about homosexual sex), although I agree with theorists who argue that queer should retain something of its "gossamer attachment to sexuality" (Freccero, "Queer Times" 489).[32] Queer is at its broadest a process of critiquing norms and binaries especially as they are deployed in power-knowledge systems, such as the body for Judith Butler. For Morton, queer ecology is about imagining multiple intimacies, celebrating rather than abjecting the fluid boundaries between human and non-human,[33] as complex meshes or networks rather than straight lines. More specifically, Catriona Sandilands has proposed that queer ecology is firstly about recognizing that "the biotic is not outside the social" and "it is a dangerous fiction to theorize an ecological ethics as if it were" ("Queering" 457). The further step that makes it queer, for Sandilands, is to understand the "implications of culture in ecosystemic relations [...] as inseparable from complex articulations of power, including but not limited to hetero-normativity" (471). An ethical eco-

---

[31] This work is nevertheless important as a base from which to challenge the oppressive instrumentalization of nature to justify heteronormativity. Potentially the most politically influential work in this vein is being done by biologists (Fausto-Sterling, Bagemihl). Katie Hogan makes a powerful case for the need to build coalitions of the excluded around a new, queer, nature.

[32] Queer theory, like ecocriticism, has its "theory wars" (see, e.g., Duggan) in which the stakes are also the questioning of theory's commitment to actual, situated, problems. Queer, Gay, Lesbian, LGBT, etc., all continue to be contested terms as definitions of academic areas of study as well as in activist communities. Gay and Lesbian Studies, and later queer theory, were challenged by feminists and lesbians as being implicitly gay-male-centered (Jeffreys). Queer theory is seen by some as elitist, classist (Taylor) and neglectful of material situated concerns of individuals (Escoffier, Morton), and of the specificity of oppressed sexualities, including those in non Anglophone countries (e.g., Pratt for France). See Butler, 223-42, for a brilliantly qualified defence of the material usefulness of an evolving, necessarily contested queer. Giffney provides a succinct overview of the evolution and state of "the 'q' word" through 2009.

[33] The growing body of scholarship on the "posthuman" should be mentioned here, which calls into question the boundaries policed by the nature-culture divide, or "rather to understand the materializing effects of particular ways of drawing boundaries between 'humans' and 'non-humans' (Barad, 123-24). On queer posthumanism, see also Patricia MacCormack.

logy is a queer ecology in that it refuses the inside-outside, or human-nature, opposition in favor of a model of interrelatedness. Again, we come back to the idea that to value culture is to give more power to ecocriticism as a theory and as a political praxis.

This idea, *mutatis mutandis*, also appears in queer historiography. Early modernists Carla Freccero, Madhavi Menon and Jonathan Goldberg, variously argue that apprehending early modernity can make our historiographic practices queer. Menon and Goldberg have called for a "homohistory" (1609). This is not a history of homosexuality as practiced for example by David Halperin, but a historiography of sameness, that resists difference and the pressures of the modernity teleology: a Latourian move, arguably.[34] Freccero argues throughout her recent work for a "return to questions of subjectivity and desire," challenging boundaries between "the world considered as an object of social scientific study and the world considered as infused with passional attachment, fantasy" ("Queer Times" 491). Queer is not so much a substantive, for Freccero, as a process of what she thinks of as ethical affective haunting, "an openness to being haunted, even inhabited" by the ethical insistence of the past (*Queer/Early* 79). Again, this is an anti-essentialist move that explicitly places affect firmly *within* any notion we might have of the world out there. A queer ecology, then, not only inquires into the way "the natural has been produced by [...] heteronormativity" (Dinshaw 77), but — sharing modalities with post-

[34] Goldberg, Menon, and Freccero have variously questioned the logic of teleology or historicity themselves, defending unhistoricism, anachronistic affinity and the presentist stakes of early modern criticism. Nardizzi, Guy-Bray and Stockton likewise argue against the "tyranny of historicism" (1). For a critique of unhistoricism, see, e.g., Traub. There is increasing concern about its perceived omissions (ethical engagement with the future, specificity of non-heterosexualities, contributions of other scholars to queer history), although Freccero is explicitly concerned with ethical futurity. A very helpful summary of queer and gay-lesbian historiographic debates is provided by Ferguson (1-54), also working on the French Renaissance, who describes his own project as inverse to that of Goldberg, Menon, and Freccero, "towards a historicisation of queer [...] from the conviction that queer still has much that it might tell us about the past" (53).

humanism — it interpolates us to find our sameness with nature, the biotic in the social and the social in the biotic.[35]

Freccero, like many queer theorists who conceive of their work as distinct from histories of homosexuality, is interested in textuality, in particular the specificity of "texts that flaunt their status as fictional, imaginative, or literary" (*Queer/Early* 51).[36] Before concluding, I would like to take up this focus on the literary to address the question of its status within ecocriticism. Ultimately, what this early modern/queer/French articulation might gesture towards within ecocriticism is the idea that we neglect the human, and this includes the humanities, at our peril. This is true even, or perhaps especially, in environmentally-oriented humanities, which are rightly concerned with the future of the non-human biosphere. Some ecocritics have shown ambivalence about the world of books inasmuch as it mediates nature; in their academic work, they have focused almost entirely on nature writing, or poetry seen as evidencing a particularly recuperative, even ecstatic, relation with nature. This somewhat guilty relationship[37] with books and representation

[35] I have here left undeveloped the specific articulations between the queer and the French. Francophone queer theory shares with *écocritique* (and French feminism) an uncomfortable relation to Anglophone academia, a dual sense that queer theory in English has something urgent to do in France, but that a home-grown French queer theory should not simply define itself by importing those notions. Leading French scholars of *le queer* include Eribon, Fassin, Bourcier and Deschamps, whose work dialogues with but distinguishes itself from Anglophone queer theory. Preciado is interested in going beyond national boundaries in thinking queerly, and Freccero proposes an always-already Frenchness to queer theory, and vice-versa (*Queer/Early* 19). Indeed, Freccero's book provides a sustained articulation between the queer, the early modern, and the French, a model which is eminently suited to interrogate notions of the natural.

[36] She proposes that "all textuality, when subjected to close reading, can be said to be queer" (*Queer/Early* 5). This relies on a broad notion of queer as ideologically perverse, rather than the more direct conflation of queer with eroticism proposed by Nardizzi, Guy-Bray, and Stockton who, following Barthes's notion of the pleasure of the text (again, the Frenchness of queer theory) liken "the backward gaze of scholarship to the backward gaze of cruising [...] a coming-together of textual and sexual pleasures" (4).

[37] The conflicted position of some ecocritics, caught between commitment to the biosphere and academic commitments to studying literature, can be seen in the evocation of Thoreau in the title for ASLE's 2003 conference, "'the solid

has also, understandably, led to demands for a more interdisciplinarity approach to the environment, for example Phillips (2003), or Glen Love who believes that the very nature(s) of ecocriticism "fairly urges its practitioners into interdisciplinarity, into science" (2003). In some of these discussions on ecocriticism and interdisciplinarity, there can be an implicit devaluing of literary studies.[38] (The distrust goes both ways, of course, since ecocritics influenced, for example, by deep ecology have often been simplistically skeptical of all scientific forms of knowledge.)[39] This might not be feasible or even desirable for many of us. First, it assumes that the sciences access a more direct way of knowing reality: Science Studies (a field in which Latour is often cited) show this to be highly questionable. Second, overly-hasty attempts to engage in other disciplines can lead to tokenistic scholarship.[40] Why should those of us trained in literary analysis not continue to pay attention to the specificity of what literature *does* in the cultural and historical realms, with — of course — rigorous awareness of our own "hauntings," positionality and privileges? A more fully humanized ecocriticism, which

---

earth! the actual world!' Sea-City-Pond-Garden. On the other hand, Bate (2000) argues for the urgency of "ecopoetics."

[38] I do agree with Phillips that literature-oriented ecocritics, if they do engage the science of ecology in attempts to be interdisciplinary, tend to do so in a tropist way (ix, 142-43); we should not, though, throw the literature out with the bathwater.

[39] One of the *ur*-essays for deep ecology's characterization of science as part of an increasingly anti-nature attitude in Judeo-Christian norms is L. White. Garrard, 20-23, has a summary of deep ecology and its characterization of scientific ecology. Heise ("Hitchhiker") and Love (*Practical*) provide good overviews of ecocriticism's fraught relations with science in the context of the "science wars" and the Sokal affair during which ecocriticism came to maturity. Love makes a sustained case for the need for greater understanding and reconciliation between science and the humanities; he describes his own shifting position away from "aggressive anti-anthropocentrism" (6) towards an understanding of humanity and nature as a system.

[40] Egan's otherwise compelling reading of a world of correspondences between text and environment in Shakespeare, for example, is arguably weakened rather than strengthened by his evocation of "the new perspectives offered by holograms, fractals, and genetics" (26). Similarly for SueEllen Campbell's cogent discussion of ecology and theory, which turns to New Age quantum physics in her discussion of observer subjectivity (129).

celebrates the coupling of nature and culture rather than insisting on their separation, would be less inclined to distance itself from textual representation. And paradoxically, if we stop insisting on how interdisciplinary we are, we might just find ourselves being interdisciplinary in new and unexpected ways. Unless, of course, we have the time and/or inclination truly to transform ourselves into scholars who analyze something other than literature, culture, or history, there is no reason why we should not continue to do what we do best, in a spirit of openness to listening to others from different fields. (Such willingness to listen to scientific ecologists, for example, would probably lead ecocritics to question our instrumentalizing of vague notions of interrelatedness.) I am here defending a certain notion of the literary, not in the terms of New Criticism or "great books with great values" ideologies, but as a category that matters, socially and materially, into which we continue to project individual and collective desires, questions, and uncertainties. Literature will certainly not save the biosphere by itself, but ecocriticism necessarily plays out in the tensions between scientific and literary discourses, and it should continue to interrogate both — as well as the tension itself.

This paper is a suggestion for possible articulations or conversations, some of which are already happening. We now have an emerging early modern ecocriticism, the beginnings of a specifically French *éco-pensée* that is not simply an exporting of English ideas into France, and queer ecology. One might conclude that it is no longer necessary to explicitly challenge ecocriticism to expand its boundaries: it seems to be doing so successfully. What I have suggested here, however, is a multi-dimensional articulation of these three fields into an approach that is *simultaneously* queer, French, and early modern. In particular, such an articulation can help ecocriticism get over the nature-culture binary that has led to critical and philosophical impasse. In different ways, early modern culture, French *éco-pensée*, and queer theory, all call for a hybrid understanding of humanity's position within nature (or nature-culture) and all the more so when taken together. In other words, they share a commitment to understanding the role of the cultural, including the literary, in the natural. I believe this hybrid nature to be more productive and helpful — in praxis, not "just" in theory — than the notion of a pure, pre-cultural nature which is absolutely opposed to the human and the social. Perhaps by accepting just how strong are the links between nature and culture, we can finally realize, as Simon Schama suggests, “just how much we stand to lose” (14) in the face of real ecocatastrophe. A new environmental ethic might turn out to be queer, French, and not new at all.

Works Cited

Adamson, Joni, Mei Mei Evans, and Rachel Stein, eds. *Environmental Justice Reader: Politics, Poetics, and Pedagogy*. Tucson: University of Arizona Press, 2002.

Alaimo, Stacy. *Undomesticated Ground: Recasting Nature as Feminist Space*. Ithaca: Cornell University Press, 2000.

Bagemihl, Bruce. *Biological Exuberance: Animal Homosexuality and Natural Diversity*. New York: St. Martin’s Press, 1999.

Barad, Karen. “Nature’s Queer Performativity.” *Qui parle* 19 (2011): 122-58.

Bate, Jonathan. *Romantic Ecology: Wordsworth and the Environmental Tradition*. London: Routledge, 1991.

______. *The Song of the Earth*. Cambridge, MA:Harvard University Press, 2000.

Baudrillard, Jean. *L’Illusion de la fin*. Paris: Galilée, 1992.

Blair, Ann. *The Theater of Nature: Jean Bodin and Renaissance Science*. Princeton: Princeton University Press, 1997.

Blanc, Nathalie. “Vers une esthétique environnementale? Regards sur un colloque.” *RACAR* 35 (2010): 11-21.

______, Denis Chartier and Thomas Pughe. “Littérature et écologie. Vers une écopoétique (Introduction).” *Écologie & politique* 36 (2008): 17-28.

Borlik, Todd. *Ecocriticism and Early Modern English Literature: Green Pastures*. New York: Routledge, 2011.

Bourcier, Marie-Helene. “Le ‘nous’ du zoo.” *Q comme Queer. Les séminaires Q du Zoo*. Lille: Cahiers Gai Kitsch Camp, 1998.

Branch, Michael. *Reading the Earth: New Directions in the Study of Literature and Environment*. Moscow, ID : University of Idaho Press, 1998.

Brown, Marshall. "Periods and Resistances." Introduction. *Modern Language Quarterly* 62 (2001): 309-16.

Buell, Lawrence. "Ecocriticism: Some Emerging Trends." *Qui Parle* 19 (2011): 87-115.

______. *The Environmental Imagination: Thoreau, Nature Writing, andthe Formation of American Culture*. Cambridge, MA: Harvard University Press, 1995.

Butler, Judith. *Bodies That Matter: On the Discursive Limits of 'Sex'*. New York: Routledge, 1993.

Campbell, SueEllen. "The Land and Language of Desire: Where Deep Ecology and Post Structuralism Meet." *Western American Literature* 24 (1989): 199-211.

Castree, Noel, and Bruce Braun. *Social Nature: Theory, Practice and Politics*. Hoboken: Wiley-Blackwell, 2001.

Cave, Terence. *The Cornucopian Text: Problems of Writing in the French Renaissance*. Oxford: Oxford University Press, 1979.

Cohen, Michael. "Blues in the Green: Ecocriticism Under Critique." *Environmental History* 9 (2004): 9-36.

Cresswell, Tim. *Place: A Short Introduction.* Oxford: Blackwell, 2004.

Cronon, William. "The Trouble with Wilderness; or, Getting Back to the Wrong Nature." *Uncommon Ground: Rethinking the Human Place in Nature*. Ed. William Cronon. New York: W. W. Norton & Co., 1995. 69-90.

Cusset, Francois. *French Theory, Foucault, Derrida, Deleuze & Cie et les mutations de la vie intellectuelle aux Etats-Unis*. Paris: Editions de la Découverte, 2003.

Desblache, Lucile. "Introduction: profil d'une éco-littérature." *Esprit Créateur* 46 (Summer 2006): 1-4.

Deschamps, Catherine. "Intro-mission." *Q comme Queer. Les séminaires Q du Zoo.* Lille: Cahiers Gai Kitsch Camp, 1998.

Descola, Philippe. *Par-delà nature et culture*. Paris: Gallimard, 2005.

Dinshaw, Carolyn. "Chaucer's Queer Touches / A Queer Touches Chaucer." *Exemplaria* 7 (1995): 75-92.

Dodson, Katrina. "Introduction. Eco/Critical Entanglements." *Qui parle* 19 (2011): 1-21.

Duggan, Lisa. "The Theory Wars, or, Who's Afraid of Judith Butler?" *Journal of Women's History* 10 (1998): 9-19.

Egan, Gabriel. *Green Shakespeare: From Ecopolitics to Ecocriticism.* New York: Routledge, 2006.

Eribon, Didier. *Réflexions sur la question gay*. Paris: Fayard, 1999.

Estok, Simon. "Bridging the Great Divide: Ecocritical Theory and the Great Unwashed." *ESC: English Studies in Canada* 31 (2005): 197-209.

______. "Theorizing in a Space of Ambivalent Openness: Ecocriticism and Ecophobia." *ISLE: Interdisciplinary Studies in Literature and Environment* 16 (2009): 203-25.

Evans, Mei Mei. "'Nature' and Environmental Justice." *The Environmental Justice Reader*, ed. Adamson et al. 181-93.

Evernden, Neil. *The Social Creation of Nature*. Baltimore: Johns Hopkins University Press, 1992.

Fassin, Eric. *L'Inversion de la question homosexuelle*. Paris: Éditions Amsterdam, 2005.

Fausto-Sterling, Anne. *Sexing the Body: Gender Politics and the Construction of Sexuality*. New York: Basic Books, 2000.

Feder, Helena. "Rethinking Multiculturalism: Theory and Nonhuman Cultures." *ISLE: Interdisciplinary Studies in Literature and Environment* 17 (2010): 775-77.

Ferguson, Gary. "Introduction." *Queer (Re)Readings in the French Renaissance. Homosexuality, Gender, Culture*. Burlington, VT: Ashgate, 2008. 1-55.

Ferry, Luc. *Le Nouvel Ordre écologique. L'Arbre, l'Animal et l'Homme.* Paris: Grasset, 1992.

Fitter, Chris. *Poetry, Space, Landscape: Toward a New Theory*. Cambridge: Cambridge University Press, 1995.

Frank, Adam. *The Constant Fire: Beyond the Science vs. Religion Debate.* Berkeley: University of California Press, 2009.

Freccero, Carla. *Queer/Early/Modern.* Durham: Duke University Press, 2006.

______. "Queer Times." *South Atlantic Quarterly* 106 (2007): 485-94.

Gaard, Greta. "Toward a Queer Ecofeminism." *Hypatia* 12 (1997): 114-37.

______, and Patrick D. Murphy. "Introduction." *Ecofeminist Literary Criticism: Theory, Interpretation, Pedagogy.* Urbana: University of Illinois Press, 1998. 1-13.

Garrard, Greg. *Ecocriticism.* New York: Routledge: 2004.

_______. "Heidegger Nazism Ecocriticism." *Interdisciplinary Studies in Literature and Environment* 17 (2010): 251-71.

Glotfelty, Cheryl, and Harold Fromm, eds. *The Ecocriticism Reader.* Athens, GA: University of Georgia Press, 1996.

Goldberg, Jonathan, and Madhavi Menon. "Queering History." *PMLA* 120 (2005): 1608–17.

Grafton, Anthony. *Defenders of the Text: The Traditions of Scholarship in an Age of Science, 1450-1800.* Cambridge, MA: Harvard University Press, 1991.

Greene, Thomas M. *The Light in Troy: Imitation and Discovery in Renaissance Poetry.* New Haven: Yale University Press, 1982.

Haraway, Donna. *Simians, Cyborgs, and Women: The Reinvention of Nature.* New York: Routledge, 1991.

Heise, Ursula K. "Ecocriticism and the Transnational Turn in American Studies." *American Literary History* 20 (2008): 381-404.

______. "The Hitchhiker's Guide to Ecocriticism." *PMLA* 121 (2006): 503-16.

Hiltner, Ken. *What Else is Pastoral? Renaissance Literature and the Environment.* Ithaca: Cornell University Press, 2011.

Hogan, Katie. "Undoing Nature: Coalition Building as Queer Environmentalism." *Queer Ecologies: Sex, Nature, Politics, Desire.* Ed. Catriona Mortimer-Sandilands and Bruce Erickson. Bloomington: Indiana University Press, 2010. 231-53.

Huggan, Graham, and Helen Tiffin. *Postcolonial Ecocriticism: Literature, Animals, Environment.* New York: Routledge, 2010.

Jeanneret, Michel. *Perpetuum Mobile. Métamorphoses des corps et des œuvres de Vinci à Montaigne*. Paris: Macula, 2000.

Jeffreys, Sheila. *Unpacking Queer Politics: A Lesbian Feminist Perspective.* Oxford: Blackwell, 2003.

Johns-Putra, Adeline. "Ecocriticism, Genre, and Climate Change: Reading the Utopian Vision of Kim Stanley Robinson's Science in the Capital Trilogy." *English Studies* 91 (2010): 744-60.

Kroeber, Karl. *Ecological Literary Criticism: Romantic Imagining and the Biology of Mind.* New York: Columbia University Press, 1994.

Larrère, Catherine. "Ethiques de l'environnement." *Multitudes* 24 (2006): 75-84.

______. *Les philosophies de l'environnement.* Paris: Presses Universitaires de France, 1997.

Latour, Bruno. *Nous n'avons jamais été modernes*. Paris: La Découverte & Syros, 1997.

______. *Politiques de la natur. Comment faire entrer les sciences en démocratie*. Paris: La Découverte & Syros, 1999.

Laughlin, Kathleen, Julie Gallagher, Dorothy Sue Cobble, Eileen Boris, Premilla Nadasen, Stephanie Gilmore and Leandra Zarnow. "Is it Time To Jump Ship? Historians Rethink the Wave Metaphor." *Feminist Formations* 22 (2010): 76-135.

Love, Glen. *Practical Ecocriticism: Literature, Biology, and the Environment.* Charlottesville: University Press of Virginia, 2003.

______. "Revaluing Nature: Toward an Ecological Criticism." *The Ecocriticism Reader: Landmarks in Literary Ecology*. Ed. Cheryll Glotfelty and Harold Fromm. Athens, GA: University of Georgia Press, 1996. 225-40.

Lovejoy, Arthur O.. *The Great Chain of Being: A Study of the History of an Idea.* Cambridge, MA: Harvard University Press, 1936.

MacCormack, Carol, and Marilyn Strathern. *Nature, Culture, and Gender.* New York: Cambridge University Press, 1980.

MacCormack, Patricia. "Queer Posthumanism: Cyborgs, Animals, Monsters, Perverts." *The Ashgate Research Companion to Queer Theory.* Burlington, VT: Ashgate, 2009. 111-26.

McGrath, Alister. *A Scientific Theology.* Vol. 1: *Nature.* Edinburgh: T & T Clark, 2001.

McLaughlin, Andrew. *Regarding Nature: Industrialism and Deep Ecology.* Albany: State University of New York Press, 1993.

Maclean, Ian. "Natural and Preternatural in Renaissance Philosophy and Medicine." *Studies in History and Philosophy of Science* 31 (2000): 331-42.

Marcus, Leah. "Renaissance/Early Modern Studies." *Redrawing the Boundaries: The Transformation of English and American Literary Studies.* Ed. Stephen Greenblatt and Giles Gunn. New York: Modern Language Association, 1992. 41-63.

Merchant, Carolyn. *The Death of Nature: Women, Ecology and the Scientific Revolution.* San Francisco: Harper & Row, 1980.

Morton, Donald. "Changing the Terms: (Virtual) Desire and (Actual) Reality." *The Material Queer: A LesBiGay Cultural Studies Reader.* Ed. Donald Morton. Boulder: Westview, 1996. 1-33.

Morton, Timothy. *Ecology Without Nature: Rethinking Environmental Aesthetics.* Cambridge, MA: Harvard University Press, 2007.

______. "Queer Ecology." *PMLA* 125 (2010): 273-82.

Murphy, Patrick. *Literature, Nature and Other: Ecofeminist Critiques.* Albany: State University of New York Press, 1995.

Myers, Jeffrey. *Converging Stories: Race, Ecology, and Environmental Justice in American Literature.* Athens, GA: University of Georgia Press, 2005.

Naess, Arne. "The Shallow and the Deep, Long-Range Ecology Movement: A Summary." *Inquiry* 16 (1973): 95-100.

Nardizzi, Vin, Stephen Guy-Bray and Will Stockton, eds. *Queer Renaissance Historiography: Backward Gaze.* Burlington: Ashgate, 2009.

Oppermann, Serpil. "Ecocriticism's Phobic Relations with Theory." *ISLE: Interdisciplinary Studies in Literature and Environment* 17 (2010): 768-70.

______. "Theorizing Ecocriticism: Toward a Postmodern Ecocritical Practice." *ISLE: Interdisciplinary Studies in Literature and Environment*, 13 (2006): 103-28.

Outka, Paul. *Race and Nature from Transcendentalism to the Harlem Renaissance*. New York: Palgrave Macmillan, 2008.

Phillips, Dana. *The Truth of Ecology: Nature, Culture and Literature in America.* Oxford: Oxford University Press, 2003.

Posthumus, Stéphanie. "Translating Ecocriticism: Dialoguing with Michel Serres." *Reconstruction: Studies in Contemporary Culture* 7 (2007). <http://reconstruction.eserver.org/072/posthumus.shtml>.

Pratt, Murray. "Post-Queer and Beyond the PaCS." *In a Queer Place.* Ed. Kate Chedgzoy, Emma Francis and Murray Pratt. Burlington, VT: Ashgate, 2002. 117-206.

Preciado, Beatriz. "Il faut queeriser l'université." *Queer: repenser les identités. Revue Rue Descartes* 40. Paris: Presses Universitaires de France, 2003. 79-83.

Pughe, Tom. "Réinventer la nature: vers une éco-poétique." *Études anglaises* 58 (2005): 68-81.

Robisch, S.K. "The Woodshed: A Response to 'Ecocriticism and Ecophobia'." *Interdisciplinary Studies in Literature and Environment* 16 (2009): 697-708.

Roger, Alain. *Court traité de paysage*. Paris: Gallimard, 1997.

Rogers, Raymond. *Nature and the Crisis of Modernity: A Critique of Contemporary Discourse on Managing the Earth.* New York: Black Rose Books, 1994.

Rueckert, William. "Literature and Ecology: An Experiment in Ecocriticism." *The Iowa Review* 9 (1978): 71-86.

Sagar, Keith. *Literature and the Crime Against Nature: From Homer to Hughes.* London: Chaucer, 2005.

Sampon-Nicolas, Annette. "Pour la nature: le cours de littérature prend l'air." *The French Review* 74 (2001): 1218-28.

Sanders, Scott Russell. "Speaking a Word for Nature." *The Ecocriticism Reader: Landmarks in Literary Ecology*. Ed. Cheryll Glotfelty and Harold Fromm. Athens, GA: University of Georgia Press, 1996. 182-95.

Sandilands, Catriona. "Lavender's Green? Some Thoughts on Queer(y)ing Environmental Politics." *UnderCurrents: Critical Environmental Studies* 6 (1994): 20-4.

______, or Mortimer-Sandilands, Catriona. "Queering Ecocritical Studies." *Cultural Studies* 22 (2008): 455-76.

______, or Mortimer-Sandilands, Catriona, and Bruce Erickson, eds. *Queer Ecologies: Sex, Nature, Politics, Desire.* Bloomington: Indiana University Press, 2010.

Schama, Simon. *Landscape and Memory*. New York: Knopf, 1995.

Serres, Michel. *Le Contrat naturel.* Paris: Bourin, 1990.

Slocombe, Will. "Littered with Meaning: The Problem of Sign Pollution in Postmodern, Post-Structuralist and Ecocritical Thought." *Textual Practice* 19 (2005): 493-508.

Slovic, Scott. *Seeking Awareness in American Nature Writing.* Salt Lake City: University of Utah Press, 1992.

______, ed. "Special Forum on Ecocriticism and Theory." *ISLE: Interdisciplinary Studies in Literature and Environment* 17 (2010): 754-99.

Soper, Kate. *What Is Nature? Culture, Politics, and the Non-Human.* Oxford: Blackwell, 1995.

Sturgeon, Noël. *Ecofeminist Natures: Race, Gender, Feminist Theory and Political Action.* London: Routledge, 1997.

Taylor, Yvette. "Queer, but Classless?" *The Ashgate Research Companion to Queer Theory.* Ed. Noreen Giffney and Michael O'Rourke. Burlington, VT: Ashgate, 2009. 199-218.

Theis, Jeffrey. *Writing the Forest in Early Modern England: A Sylvan Pastoral Nation*. Pittsburgh: Duquesne University Press, 2009.

Traub, Valerie. "The Present Future of Lesbian Historiography." *Blackwell Companion to LGBT/Q Studies*. Ed. George Haggerty and Molly McGarry. London: Blackwell, 2007. 124-45.

Vandenberge, Frédéric. "The Nature of Culture. Towards a Realist Phenomenology of Material, Animal and Human Nature." *Journal for the Theory of Social Behavior* 33 (2003): 461-75.

Vital, Anthony. "Toward an African Ecocriticism: Postcolonialism, Ecology and Life & Times of Michael K." *Research in African Literatures* 39 (2008): 87-121.

Wang, Ning. "Toward a Literary Environmental Ethics: A Reflection on Ecocriticism." *Neohelicon* 36 (2009): 289-98.

Watson, Robert. *Back to Nature: The Green and the Real in the Late Renaissance*. Philadelphia: University of Pennsylvania Press, 2006.

Westphal, Bertrand. *La Géocritique. Réel, fiction, espace*. Paris: Minuit, 2007.

White, Lynn, Jr. "The Historical Roots of Our Ecologic Crisis." *Science* 155 (1967): 1203-07.

White, Richard. *The Organic Machine*. New York: Hill and Wang, 1995.

Whiteside, Kerry. *Divided Natures: French Contributions to Political Ecology*. Cambridge, MA and London: MIT Press, 2002.

Williams, Raymond. *Keywords: A Vocabulary of Culture and Society*. 2nd ed. New York: Oxford University Press, 1983.

Wilson, Alexander. *The Culture of Nature: North American Landscape from Disney to the Exxon Valdez*. Toronto: Between the Lines, 1991.

Christophe Ippolito

The Georgia Institute of Technology

Discours mythiques et écocritiques en dialogue: l'exemple d'Alexandrie d'Égypte

Cet article tente de mesurer comment l'héritage théorique et méthodique des études sur le mythe dans les années 1970 et 1980 peut jouer un rôle dans le contexte actuel des études écocritiques. L'exemple choisi est celui d'une ville fondée dans un environnement naturel particulier, Alexandrie, et en cela l'article met en question toute division stricte entre environnement "naturel" et "non-naturel."

Dans les années 1970 et 1980 en France, il était pratiquement impossible aux critiques, en particulier en littérature comparée, d'éviter le discours sur le mythe; ce mouvement a perdu de son ampleur au début des années 2000, et cela au moment où l'écocritique et le discours sur l'environnement en littérature ont commencé à se développer de façon autonome. Compte tenu de l'espace imparti, sans rentrer dans le détail des textes littéraires, on voudrait montrer comment le discours théorique d'alors sur le mythe, dont on revisitera ici les manifestations structuralistes, peut contribuer à servir et codifier le discours sur l'environnement en littérature, et notamment les représentions de l'univers urbain, et comment ces deux discours peuvent s'articuler. Dans le même temps, on appréciera, dans le prolongement des travaux de Timothy Morton sur la critique des perspectives écocritiques conservatrices (161), comment ce que l'idée reçue voit comme le contraire de l'environnement physique (la ville) peut paradoxalement déconstruire et sublimer cet environnement dans le cadre mythique. L'exemple choisi sera celui d'Alexandrie d'Égypte,[1] à travers les représentations qui en ont été faites entre 1798 et 1952, de l'invasion de Bonaparte à l'avènement de Nasser (qui a profondément changé la ville où il est né en l'orientant vers l'intérieur du pays, et non plus *ad Aegyptum*).

Parler du mythe d'Alexandrie, c'est reconnaître que ce mythe, dans ses éléments (la ville méditerranéenne aux confins du delta du Nil et du désert, la fondation, la métropole antique, la bibliothèque, la sensualité, l'image historique, la décadence...), a permis l'éclosion d'une littérature européenne et notamment de fictions entre 1798 et 1956, dans la mesure où, comme le jugeait Lévi-Strauss (131-43), les mythes meurent en créant des romans. Étudier les conditions de possibilités de la transformation de la ville en mythe dans la littérature moderne et contemporaine reviendrait ainsi à donner à l'histoire d'une création mythique continuée de l'Antiquité à aujourd'hui une introduction, ou du moins un mode d'emploi possible. Alexandrie, dans la littérature européenne moderne, serait la projection mythique d'un schème, à savoir l'idée que la civilisation européenne méditerranéenne se fait d'elle-même, dans un lieu privilégié de l'espace physique égyptien. Cette ville serait représentée comme l'archétype de cette civilisation: ville européenne exemplaire, adéquate à sa situation dans un espace physique donné.

Bien entendu, dans le cas d'Alexandrie, cette image primordiale est fausse sur le plan de la vérité historique. Il n'empêche qu'elle est reconstituée dans certains romans ou poèmes comme une projection des hantises occidentales (incompréhension occidentale de l'"Autre" arabe, et surtout, en somme, autojustification); car d'un point de vue idéologique, c'est un moyen de justifier une situation coloniale caractéristique d'un certain discours orientaliste par lequel les Européens essayèrent de définir le Moyen Orient comme l'Autre exotique et stéréotypé de l'Europe. La constitution du mythe d'Alexandrie serait alors une tentative de rechercher le Même sous l'Autre, et une réappropriation conçue pour l'intégrer à un univers européen. Dans cette perspective, l'archétype, à mi-chemin entre le schème et les images perçues dans l'environnement, est la substance symbolique du schème, "le noumène de l'image que l'intuition perçoit" (Jung 411). Les archétypes ont une

---

[1] Je remercie le ministère français des affaires étrangères et le CEDEJ, et le gouvernement égyptien et son service des antiquités, de leur aide pour la partie de ce travail menée au Caire et à Alexandrie.

fonction particulière: ils permettent de structurer l'imaginaire. Plusieurs images d'Alexandrie constitueraient ainsi une idée d'Alexandrie, et en retour l'idée d'Alexandrie ainsi élaborée constituerait plusieurs images. Ce double mouvement est caractéristique de la création mythique comme continuum. Allant ainsi au-delà du symbole, l'archétype nous donne une base de données non pas seulement polyvalente, mais de valeur universelle et qui gouverne et structure dans l'espace le cycle temporel de l'éternel retour des apogées et des déclins qui caractérise Alexandrie (un million d'habitants sous l'empire romain, 7000 en 1798, cinq millions aujourd'hui).

Dès lors, on peut assimiler Alexandrie dans l'imaginaire européen à un archétype qui serait "le stade préliminaire, la zone matricielle de l'idée" (Jung 411) que se font de la ville en général les auteurs européens qui écrivent sur elle. Pour chaque écrivain, "bien loin de primer l'image, l'idée ne serait que l'engagement pragmatique de l'archétype imaginaire, dans un contexte historique et épistémologique donné" (Durand 62). Dès lors, il est évident qu'on a des représentations différentes à partir d'une idée originelle; Alexandrie, mythe littéraire, peut être perçu de façons diverses: soit décadente, soit conquérante, soit emprise dans la réalité de sa situation topologique, soit imaginaire, soit encore miroir du Paris fin-de-siècle, soit enfin image privilégiée de la ville méditerranéenne au même titre que Beyrouth, Marseille, ou Naples.

L'Alexandrie littéraire est avant tout un espace imaginaire organisé par son mythe, dans la mesure où, comme l'écrit Jean Starobinski, "[...] nous voyons se dessiner une tâche critique qui ne se limiterait pas à l'analyse de l'univers imaginé, mais qui observerait la puissance imaginante dans sa situation relative au sein du contexte humain où elle surgi" (194-95). La position bien connue de Gilbert Durand sur la structure profonde comme réseau de forces en tension se prête particulièrement à la description de l'environnement urbain et proprement physique de l'Alexandrie mythique (8); et le mythe est constitué de mythèmes qui, telle la relation d'incorporation du passé au présent, se caractérisent par cette tension entre deux éléments souvent contraires. L'hétérogénéité même des diverses composantes alexandrines, du cosmopolitisme à la position géographique tangentielle, renforce cette idée de tension.

Cependant, il n'existe pas de récit modèle aisément identifiable à propos d'Alexandrie (si l'on met de côté, comme concernant seulement un aspect du mythe, les récits de fondation). Mais "il n'est peut-être pas assuré que tout mythe littéraire indique un récit" (Trousson 174). Raymond Trousson rappelle pertinemment que le mythe de situation, celui de l'Éden par exemple, n'existe qu'en fonction d'une chaîne de *données* narratives (c'est nous qui soulignons), d'une syntaxe du récit. Cette chaîne de données s'organise ici comme la projection de l'archétype d'Alexandrie comme emblème de la civilisation européenne dans certaines œuvres de la littérature européenne. Cette projection est une construction à un double titre.

Tout d'abord, c'est une construction idéologique qui repose sur les fondements suivants: Alexandrie fut considérée comme un poste avancé de l'Occident de 1798 à 1952; ville où vivent de nombreux Européens dans une atmosphère de cosmopolitisme, sa transformation en mythe sert à justifier la présence de cette communauté. Les références au passé antique sont là, il ne reste qu'à oublier le réel et à les convoquer pour faire œuvre de créateur de mythe, en cherchant notamment à travers le phénomène discursif une surdétermination sociale et historique. Et en ce sens il y a bien sûr mystification, même si cette mystification repose pour partie sur les données historiques; certaines données sont privilégiées, et donnent lieu à l'établissement d'une version tronquée et partiale des faits. Dans cette perspective, Jean-Louis Backès voulait écarter du territoire du mythe "la trop commode notion de survivance, lit de Procuste où s'élabore [...] l'image homogène que les civilisations cherchent à donner d'elles-mêmes" (24). Mais cette image mythique constituée par la civilisation occidentale et ses représentants à Alexandrie existe, et par elle, ville et espace "premier" ont partie liée.[2]

---

[2] "Lorsqu'un groupe quelconque fixe sa résidence en un point de la planète, il transfigure l'emplacement naturel et sa géographie physique en un lieu humain; il superpose à la réalité matérielle un espace et un temps qui lui sont propres, au sein desquels se déploieront les activités coutumières. Première alliance de l'homme et du monde, indispensable à la sécurité de la vie, ce fondement ontologique de la présence s'énonce sous la forme de mythes, qui justifient les rythmes de l'existence communautaire. Le corpus mythique peut revêtir des formes très variées, mais la fonction mythique est partout la même; elle intervient comme le principe de conservation de l'ordre communautaire" (Gusdorf 43).

En fait, comme le disait Pierre Brunel à propos du mythe de la métamorphose, on ne peut parler du mythe d'Alexandrie "qu'en se plaçant, non sans quelques artifices, à un point de vue synthétique" (8). Comme lui, nous n'adopterons ce terme que "comme une approximation" (8). Premièrement, le mythe peut se définir par cette fonction: appréhender une vérité qui se dérobe à la seule raison, au logos. Car le mythe d'Alexandrie est d'une certaine manière un principe explicatif. Il a été "fabriqué dans un second temps pour justifier un rite, un usage" (Backès 22), par certains écrivains européens qui projettent leurs hantises, leurs fantasmes, sur une cité européenne en Asie. Deuxièmement, le nom du mythe alexandrin renvoie à une origine, à la fondation de la ville dans la mesure où "la pensée mythique est inséparable d'une exaltation des tout premiers commencements" (Backès 22). Troisièmement, "le déplacement, ou la métonymie, [est] un trait du discours mythique" (Albouy 354). On dit ainsi "Alexandrie" pour les Alexandrins. "La ville acquiert les qualités les plus éminentes de ses habitants et se transforme en un être vivant fortement caractérisé" (Albouy 354). Par suite de ce déplacement, Alexandrie, capitale des Lagides, de la civilisation hellénistique, devient l'emblème de cette civilisation hellénistique elle-même, et pour tout dire, de la civilisation européenne. Métaphores et métonymies tirent la ville vers le mythe. Dans cette perspective, Alexandrie, comme c'est le cas de Paris selon Pierre Albouy, devient à la fois être cosmique (alliance, jonction, fiction de deux caractéristiques de l'environnement physique, mer et désert, chez Anatole France) et être individuel (conquérante, courtisane: Thaïs, également chez France). "Ce sont ces changements de plan, ce passage à un ordre de faits hétérogènes, qui constituent un des traits distinctifs du mythe" (Albouy 354). Et ici encore espace "premier" naturel et ville ont partie liée.

Sur le plan historique et sociologique, on est fondé à parler d'un mythe d'Alexandrie après Robert Ilbert, auteur d'une thèse remarquée et citée en bibliographie sur la composition sociologique de la ville dans la période retenue.[3] Au-delà, ville et texte (littéraire ou non) sont indisso-

---

[3] "[La ville] a pu devenir mythe [...]. Si Alexandrie a pu être LA ville c'est parce qu'elle représentait cet instant particulier où la mer et le port ne faisaient qu'un. [...] Et c'est un temps achevé de l'histoire. D'où aussi la force de l'image devenue imaginaire. D'où le mythe alexandrin qui a dû être, mais qui aujourd'hui n'est plus que mémoire. Ville du temps perdu qui lentement s'éteint, comme ces vieilles familles grecques qui disparaissent une à une,

lublement unis par des liens à la fois historiques (c'est historiquement dans les premières villes qu'apparaissent les premiers textes) et topologiques: la ville est le lieu où sont produits la plupart des textes, mais plus intéressant, la multiplicité des textes crée en retour une justification pour la métropole.[4] Alexandrie est fille des textes écrits sur elle et de l'espace qu'elle occupe. C'est cette hypothèse que développe Michel Butor en insistant sur la place centrale de la bibliothèque dans la ville.[5]

---

laissant chez le brocanteur leurs cartons de souvenirs que l'historien va rechercher" (Ilbert 1984, 77). Dans l'analyse sociologique, l'environnement physique revient encore.

[4] Comme le mythe de Paris, le mythe d'Alexandrie est rarement au centre d'une œuvre en littérature française, comme cela a pu être le cas en grec moderne (avec Cavafis et Tsirkas), en italien (avec Ungaretti), ou en anglais (avec Lawrence Durrell; notons qu'Ungaretti et Durrell mettent l'accent, comme Anatole France, Barrès et E. M. Forster, sur l'environnement physique: mer et désert). C'est cependant bien le cas dans *Thaïs* et *Aphrodite*. "Dans bon nombre d'autres cas, on est plutôt en présence d'une tendance au mythe" (Albouy 355). Si l'espace le permettait, on pourrait convoquer à titre d'exemple quelques textes (de Barrès par exemple) qui semblent faire du mythe d'Alexandrie une évidence partagée. L'Alexandrie mythique domine ceux qui l'habitent en poètes, et les autres, ces personnages secondaires des banquets de Louis Ménard. L'expression "mythe d'Alexandrie", bien qu'accréditée dans plusieurs récits ou études (cf. Ilbert 1984, 77), demande à être précisée, comme pour le mythe de Paris: "Quel sens revêt le mot mythe, dans ce cas? Le mythe fait partie de l'idéologie, définie comme un système de représentations constituant la représentation imaginaire des rapports réels de l'individu à la société et à son histoire. La fonction du mythe est celle-là même de l'idéologie, qui est de permettre la reproduction de ces rapports réels des hommes dans la société; le mythe exerce donc une fonction d'exaltation ou de défense, apte à aider à la reproduction des rapports des individus avec telle formation sociale" (Albouy 353). Ainsi, le mythe hellénistique de la fusion Europe-Asie serait un élément capital du mythe d'Alexandrie.

[5] "Le siège de l'autorité, ce n'est pas tellement celui du gouvernement, du chef militaire, du grand-prêtre, c'est celui des archives" (Butor 36). On connaît l'importance politique et symbolique de la Bibliothèque pour les Ptolémées, qui y plaçaient des hommes de confiance. Sous César, son incendie marque leur chute.

Mais le récit de sa fondation par Plutarque insistait d'abord sur l'environnement physique,[6] et en faisait le socle du mythe de la ville.

Michel Butor se disait surpris, lors d'une intervention au Centre culturel français de Mounira au Caire en 1986, de l'abondance des enseignes lumineuses qui sont autant de signes dans la nuit, au point même que certaines avenues ont été rebaptisées dans le dialecte cairote du nom des publicités qu'elles étalent: avenue Marlboro à Mohandessine par exemple. Les signes culturels, de fait, abondent et cette abondance rend le déchiffrage de la ville ardu et efface l'environnement naturel (la falaise du Mokattam, par exemple). La ville ne s'assimile pas à un texte prêt-à-lire, si encyclopédique soit-il. Et il n'existe pas de texte primitif pour décrire cet objet, de sorte que la ville-livre est une manière

---

[6] Plutarque (ci-après dans la traduction d'Amyot qui a eu tant d'influence dans la culture française) insiste d'abord sur le caractère prédestiné de l'étendue que représente l'île de Pharos; elle est du point de vue d'Alexandre "l'assiette du monde la plus propre, pour ce qu'il avoit en pensée de faire" (Plutarque 64); il note aussi la situation géographique particulière de cette île "un peu au-dessus de la bouche du Nil" (64), soit proprement "ad Aegyptum," aux marges de l'Egypte; c'est en quelque sorte aussi un port "naturel," "car c'est comme une langue ou une encouleure de terre assez raisonnablement large, qui separe un grand lac d'un costé, et la mer de l'autre, laquelle se va là aboutissant à un grand port" (64). La correspondance privilégiée entre étendue et espace revient dans la suite des propos de Plutarque, qui indique qu'Alexandre "commanda que promptement on lui trassast et designast la forme de la ville selon l'assiette du lieu" (64; Julien Gracq, qui avait bien compris l'importance de l'environnement physique pour une ville, a employé une formule similaire dans le titre de son livre sur Nantes, *La forme d'une ville*). Suit la description de la fondation de la ville, qui est décrite comme une alliance entre la nature et l'homme: "[à défaut de terre blanche] ilz prirent de la farine, dont ilz trasserent dessus la terre qui estoit noire, une grande enceinte courbée en figure circulaire, le rond de laquelle se terminait par le dedans en deux bases droittes de grandeur egale, qui venoyent à clorre toute la grandeur de ce pourpris en forme de manteau macedonique. Alexandre en trouva le portraict beau et y prit grand plaisir; mais soudainement une multitude infinie de grands oyseaux de toutes especes se leva du lac et de la riviere, en si grand nombre qu'ilz obscurcissoyent l'air comme eust fait une grosse nuée, et, venans à se poser en ce pourpris là, mangerent toute la farine sans qu'il y en demourast chose quelconque. Alexandre se troubla de ce presage; mais les devins lui dirent qu'il ne falloit point qu'il s'en faschast, pource que c'estoit signe qu'il bastiroit là une ville si plantureuse de tous biens, qu'elle suffiroit à nourrir toutes sortes de gens" (64-65).

de palimpseste. D'où la tendance à se reporter à des représentations littéraires que le mythe peut unifier.[7] Chaque texte qui s'ajoute hérite d'une certaine façon des éléments structurants apportés par le mythe, mais aussi par les textes successifs. Une seconde série d'éléments, plus difficiles à appréhender parce que transitoires, consisterait en des signes émis par la ville elle-même (noms des stations de tramway, par exemple). Ces signes instables, liés au collectif anonyme que représente la ville, s'apparentent à ceux du romanesque qui les réutilise et parfois les amplifie en les superposant aux signes "naturels" de l'environnement physique.

Comme le roman, comme la nature, la ville constitue un réseau à déchiffrer où se trament des destinées individuelles.[8] La ville est donc, comme le texte, un outil structurant. La construction d'une ville et celle d'un roman ont ainsi des traits communs, et des outils topologiques comparables:

> C'est sur cette stratification, sur ces diversités, voire sur ces différences de langage, qu'il [le romancier] bâtit son style, tout en conservant l'unité de sa personnalité de créateur, et l'unité (d'un autre ordre, il est vrai) de son style. (Bakhtine 119)

---

[7] Si toutefois l'on voulait trouver non plus de simples signes mais des textes (en se limitant au domaine français) sur lesquels l'imaginaire de la ville à l'œuvre dans la vision européenne puisse prendre appui, ce seraient d'abord des fictions: *Les Martyrs* de Chateaubriand, *La Tentation de saint Antoine* de Flaubert, "Désintéressement" de Barrès (dans *Sous l'œil des Barbares*), *Thaïs* d'Anatole France ou *Aphrodite* de Pierre Louÿs ; complèteraient le corpus des récits de voyage de Chateaubriand, Nerval, Du Camp, Flaubert, Fromentin, Gautier, Gobineau, Renan, Barrès, Cocteau, et des essais de Jean Grenier ou Etiemble, mais aussi des textes francophones autochtones de la première partie du vingtième siècle.

[8] "Le genre littéraire qu'est la ville peut être aisément comparé au roman. Le grand romancier étant en effet celui qui sait faire entendre la voix de ses personnages, doit donner à chacun un style particulier, son style est à proprement parler un superstyle, une intégration de ces phénomènes stylistiques qui suffiraient parfaitement à caractériser un orateur ou un poète lyrique. De même, le style d'une grande ville intègre une prodigieuse quantité de sous-styles, ceux de ses villages juxtaposés ou superposés" (Butor 37).

Ville et roman s'énoncent parfois en des termes semblables; quand Mikhaïl Bakhtine parle à propos du roman de "monde plurilingue et polyphonique" (187) ou d'une "confusion de Babel" (105), ces termes pourraient s'appliquer à la ville. Inversement, lorsque Pierre Sansot écrit que "la ville s'offre comme une totalité que nous n'abordons que par des perspectives" (51), cette observation pourrait s'appliquer au roman et à (l'observation de) la nature.

On le sait, Alexandrie se caractérise, comme toute ville, par l'apport de couches successives (architecturales, historiques, culturelles), dans le temps, mais aussi dans l'espace: île, digue, quartiers grecs, etc. *Thaïs*, le roman d'Anatole France, en particulier, témoigne d'une discontinuité de la narration qui redouble celle de la ville, parce que le temps éternel du héros Paphnuce et celui de la ville ne sont pas de la même essence, mais aussi parce que la pureté du désert de l'anachorète rentre avec ce héros dans la ville. L'objet qu'est la ville est un "palimpsest-in-progress" qui n'oublie jamais la nature qu'il a envahie et qui sans cesse se rappelle à lui: chacune de ses significations, chacun de ses signes en appellent d'autres, sans que la compréhension puisse reposer sur un sens définitivement établi. Elle n'a pas de sens définitif, et en cela c'est une nature naturante.

Ville et roman moderne ont un point commun essentiel: la spatialité organisée; l'espace construit de la métropole rencontre la structure du roman polyphonique et relativiste. Comme ce dernier, la ville se dit en plusieurs sens, s'énonce par des voix diverses, se constitue comme un dialogue, un forum. De nombreux textes de la fin du dix-neuvième siècle reflètent à la manière d'un kaléidoscope l'Alexandrie de perdition (à la fois Éden inversé, nature corrompue et serre chaude) que Flaubert décrit au début de la version définitive de *La Tentation de saint Antoine*. Discours décadent pour ville décadente, Grande Prostituée babylonienne (pour reprendre l'image biblique). La métropole et l'écriture romanesque moderne ont des liens privilégiés. Le mythe de la décadence, qui structure les forces de décomposition à l'œuvre dans l'éclatement de la ville, trouve de la sorte un répondant dans le roman (ainsi chez Louÿs). Le discours romanesque sur la ville est caractéristique du dialogisme que Bakhtine a théorisé: il ne saurait être univoque, parce qu'il intègre une multiplicité de points de vues,[9] y compris ceux qui insistent de façon critique sur l'environnement physique qui la conditionne.

Parce qu'elle est de l'ordre du palimpseste, la ville est un objet littéraire qui favorise de manière privilégiée le dialogisme, et, sur le plan de la création romanesque, l'exercice de l'intertextualité. Cependant le discours d'autrui détermine également la ville: "[...] maintenant ce milieu plurilingue de mots "étrangers" se présente au locuteur non plus comme l'objet, mais dans le cœur de l'interlocuteur, comme son fonds aperceptif, lourd de réponses et d'objections" (Bakhtine 104). La construction échappe au constructeur, si l'on veut. Dans ces conditions, recréer une ville imaginaire est un moyen de lui redonner une structure qui permet de nouveau de l'appréhender dans sa totalité. Ainsi, en distribuant des jalons sur la topologie imaginaire de la ville, on se donne des références pour expliquer sa réalité. "Le mythe urbain, quand il existe, est en partie, un écho du lieu" (Sansot 36): cette affirmation de Pierre Sansot s'applique bien à Alexandrie où le mythe est l'écho d'un lieu de légende, celui de la fondation de la ville dans la version de Plutarque. Dans ce contexte, si l'on comprend que le discours du mythe n'est pas un outil daté et peut apporter une unité et une perspective aux réflexions sur la littérature et l'environnement aujourd'hui, on comprend aussi que ce n'est qu'artificiellement qu'on peut opposer l'environnement "naturel" (presque partout transformé par l'action de l'homme) à l'environnement urbain qui en propose une limite.

---

[9] Rappelons qu'on peut parler de dialogisme selon Bakhtine dans la mesure où l'objet littéraire exige toujours une interprétation qui est "le point de convergences de voix diverses" (Bakhtine 102): "Il est entortillé, pénétré par les idées générales, les vues, les appréciations, les définitions d'autrui. Orienté sur son objet, il (le discours concret) pénètre dans ce milieu de mots étrangers agité de dialogues et tendu de mots, de jugements, d'accents étrangers, se faufile dans leurs interactions compliquées, fusionne avec les uns, se détache des autres, se croise avec les troisièmes" (Bakhtine 100).

Ouvrages cités

Albouy, Pierre. *Mythographies*. Paris: José Corti, 1976.

Backès, Jean-Louis. “Mythe et idéologie.” *Mythes, images, représentation*. Éd. Jean-Marie Grassin. Limoges: Trames, 1977 (1981). 17-25.

Bakhtine, Mikhaïl. *Esthétique et théorie du roman*. Paris: Gallimard, 1978.

Brunel, Pierre. *Le Mythe de la métamorphose*. Paris: Armand Colin, 1974.

Butor, Michel. *Répertoire V*. Paris: Minuit, 1982.

Durand, Gilbert. *Les Structures anthropologiques de l'imaginaire*. 1960. Paris: Bordas, 1969.

Flaubert, Gustave. *La Tentation de saint Antoine*. In *Œuvres complètes*. Ed. Bernard Masson. Paris: Editions du Seuil, 1964. Tome I.

France, Anatole. *Thaïs*. [1891] Paris: Collection Pourpre, s.d.

Grassin, Jean-Marie, éd. *Mythes, images, représentation*. Actes du XIVème congrès, Limoges, 1977, de la Société Française de Littérature Générale et Comparée. Limoges/Paris: Trames /Didier, 1979.

Gusdorf, Georges. *Mythe et métaphysique*. Paris: Flammarion, 1984.

Ilbert, Robert. *Alexandrie, espace et société*, 1830-1930. Thèse pour le doctorat d'état. Paris: EHESS, Janvier 1990.

_____. “Alexandrie, la mémoire et le mythe.” *Le Miroir égyptien*. Éd. Robert Ilbert et Philippe Joutard. Marseille: Éditions du Quai/ Jeanne Laffitte, 1984. 71-84.

Jung, Carl J. *Types psychologiques*. Trad. Yves Le Lay. Genève: Georg, 1986.

Lévi-Strauss, Claude. “Comment meurent les mythes.” *Science et conscience de la société. Mélanges en l'honneur de Raymond Aron*. 1:131-43. Paris: Calmann-Lévy, 1971. Réimprimé dans *Anthropologie structurale deux*. Paris: Plon, 1973.

Morton, Timothy. *Ecology without Nature: Rethinking Environmental Aesthetics*. Cambridge, MA: Harvard University Press, 2007.

Plutarque. *Les Vies des hommes illustres*. Tome VII. Trad. Jacques Amyot. Éd. Antoine Allègre. Paris: Janet et Cotelle, 1818.

Sansot, Pierre. *Poétique de la ville*. Paris: Klincksieck, 1971.

Trousson, Raymond. "Mythes, domaines et méthodes." *Mythes, images, représentation*. Éd. Jean-Marie Grassin. 1979. 171-79.

Walter Putnam

University of New Mexico
Senior Research Fellow, Faculty of Humanities, University of Johannesburg

"Please Don't Feed the Natives": Human Zoos, Colonial Desire, and Bodies on Display

The widespread practice of displaying indigenous peoples in colonial fairs contributed to the creation of a sense of radical difference between colonizer and colonized, between European and non-European, between familiar and strange. The physical proximity afforded by close-up encounters with exoticized natives promoted an ideological position of superiority among European viewers whose power and prestige were confirmed by such displays. By removing these "specimens" from their "natural" environments, Europeans created a spectacle devoid of any meaningful context. Through control of the circulation of images and ideas, colonial powers were able to confirm their own centrality and superiority in the latter nineteenth and early twentieth centuries.

> We are here in Paris; all is fine in this great city where there are so many people that the whites are packed together like corn stalks in a field. Everyone is well. The great commander, M. de Brazza, is with us. And we are happy to be here. Tell everyone that, when we come home, we will have so much to tell about all that we have seen that we will talk for months and months without stopping. (*Exposition Universelle* 177, my translation)

This letter, published in the official catalog of the 1889 Universal Exhibition in Paris, was purportedly written by Mamouka, a canoe paddler, back to his loved ones in Gabon. He was one of several hundred Africans put on display in the African villages along the Champ de Mars at the foot of the recently-constructed Eiffel Tower. His perfect French, like his perfect feelings, cannot but have been contrived. Another article from the same 1889 exhibit affirmed that the Guineans placed on exhibit would come to appreciate the glories of the French capital and return home to tell tales favorable to the French and to prepare a warm reception for future French colonists (Blanchard et al., *Le Paris noir* 29). During the latter nineteenth and early twentieth centuries, the practice of exhibiting indigenous peoples was widespread and commonly accepted. These spectacles of radical alterity placed European and North American audiences in direct contact with "specimens" (for that is the term that was commonly used) from exotic lands. The displacement of natives not as slaves or captives, not as migrants or refugees, not as workers or tourists, but for the amusement and edification of their Western audiences has a long genealogy. The history of these exhibits, like most colonial policies and practices, mirrored internal mutations in Western societies. Many of these are the same mutations that could also go toward explaining the appearance of zoos across Europe and North America in the late nineteenth century, whether for non-human or human animals. Among the most important are: rapid industrialization and urbanization of Euro-American nations following the Industrial Revolution, increased distance from wilderness and nature, changes in notions of work and leisure, shrinking of space and preoccupation with speed, and a heightened brand of cultural anxiety manifested in racial discourse, class tensions and sexual realignments.

By the outbreak of World War I, European imperialism had gained control of nearly 85% of the earth's inhabitable surface (Pieterse 76), while the resulting encounter of cultures led to what Aimé Césaire frequently termed the "colonial shock." The new colonial powers were curious about the extent and content of their domain. Human zoos served ideological and propagandist purposes by attempting to convince a skeptical metropolitan population that the colonial adventure was working and that it was worth the cost.[1] "Savages" soon become "natives" on their way to becoming "colonial subjects" and even "immigrants" according to the official scheme of cultural and political evolution. Their transformation from "objects of curiosity" to "subjects on display" followed the itinerary of racialization that informed public beliefs in the latter half of the nineteenth century. Situated at the juncture between science and show biz, these "living museums," as they were sometimes called, contributed profoundly to Western perceptions

[1] For a fully developed study of metropolitan attitudes toward France's overseas empire, consult Raoul Girardet's *L'Idée coloniale en France de 1871 à 1962*.

and stereotypes of the exotic Other. These spectacles were a fundamentally vicarious and voyeuristic phenomenon. They involved the erasure of context and the recreation of a simulacrum of authenticity whereby natives were recontextualized, removed from their original environments and plopped down in a foreign setting with the expectation that they would act "naturally."

Colonial subjects exhibited in zoos or fairs were framed as objects taken from an imagined environment and put on display for their very nature. According to the racial paradigm of the day, natives were supposed to embody nature; they were supposed to be natural, even in the most unnatural of settings. Senegalese dancers at the foot of the Eiffel Tower, Congolese pygmies at the Folies-Bergères, Ashanti warriors at the Jardin d'Acclimatation all represented denatured "specimens" of types taken out of context and reframed to suit the purposes of their European audiences. They bore an uncanny likeness to their European counterparts who, nevertheless, felt compelled to cast them as creatures of radical alterity, emerged from strange and savage lands, products of an environment that was both hostile and inviting. Natives came to embody nature itself. Indeed, "nature" and "native" share a Latin root, from *nascor, nasci*, *natus* "to be born." They were creatures of their upbringing, shaped by their environs, adapted to a world that had meaning to them. This belief in nature was largely founded on Romantic notions that were being challenged by the accelerated commerce of commodified goods and ideas. This attachment to a hypothetical place of origin relies on the belief that nature is an eternal force and presence. It has laws and is collective and plural. Nature is construed as timeless, not historical, not cultural. Such is the backdrop against which culture evolves in historical time. By framing natives in settings that construct nature as a timeless idyll, they become "stuck in time." Spectators at human zoos and ethnographic spectacles yearned for a glimpse at the past, most often conceived as timeless and atavistic. This epistemological act of peering backwards into an immutable past denies agency and subjectivity to the specimens on display. They are transformed into objects, fixed signifiers waiting to be defined by their Western conquerors. In the context of the nineteenth-century fixation on progress, natives, like nature, were left behind. They were cast as creatures, or victims, of their environments, prehistoric relics of a past that Europeans had largely escaped.

The history of representations of the other would take us far afield and into a debate that I first need to set aside before going much further. My point, simply put, is that human zoos are not pure representations in that representations can be considered mediated substitutes for something that is not present. A representation stands in for an absence. Representations are largely metaphorical tropes in that they rely on distance in attempting to bridge a gap between here and there, between then and now. Human zoos, on the other hand, were so successful because they presented the unique possibility of a seemingly unmediated, direct witnessing of the Other. They shrank distance to the space of a cage, a roped-off area, an arena or a stage, sometimes even allowing for direct physical contact to take place between observed and observer. Rather than establishing the distance of metaphor, they underscore the proximity of metonomy — and metonomy is the privileged trope of colonial discourse because it proceeds by juxtaposition and contiguity. Human zoos allow viewers to satisfy their desire for first-hand authenticity and to extrapolate about the wider world based on a visible fragment of that world placed before them. Their success relies on the cognitive and ideological conflation of environments from the imagined absence of setting to the confirmed presence of the exhibit itself. It is not by chance that humans were exhibited in zoos alongside animals, often displayed insidiously alongside primates at a time when the search for the infamous "missing link" was reaching fever pitch. The juxtaposition of species and ethnic types blurred the racial boundaries between human and non-human animals just as hierarchies were being elaborated on the basis of physical characteristics. In a taxonomic move paralleled by the ordering of nature that was taking place in zoos, representative types of exotic peoples not only stood in for larger groups of their homologues but they more importantly served to create those types in the Western imagination. Because these human zoos allowed eyewitness verification of the existence of different racial and cultural types, such exhibits also served as privileged markers of Western hegemony. Finally, I want to insist on the performative nature of their impact on Euro-American imagination as it constructed its own identity in relationship and in opposition to peoples of other racial, ethnic, geographical and cultural origins. It was this creation of an environment and a context for ideological purposes that made the human zoo phenomenon so insidious.

Over the latter half of the nineteenth century, anatomy and physiology became the privileged criteria for classifying people and placing them within a taxonomic system. The body became the site of distinctive traits that were widely studied by phrenologists who gave the aura of scientific validity to these displays by examining every specimen that came to them. The ordering of nature through a purported scientific system relied on devising families, groupings, and categories in an arboreal structure that supposed core unity and structural diversity. This representation of the natural world as a tree-like schema created a master narrative that became dogma in Europe after Darwin and Mendel had cracked the code of evolution and genetic variation. The physical body on display, often with minimal clothing, also became quickly eroticized, feeding into the anxiety about sexuality that was such an important cultural turn in late nineteenth-century Europe and North America. The physical presence of indigenous peoples who could move, dance, play sports, cook, etc. led to a striking performance of difference that made these displays powerful and memorable in the popular imagination. Their kinetic presence stood in opposition to the more static, mediated modes of representation by still images that provided a single iteration of each subject. Just as imaging technologies were evolving from still photography to motion pictures, so physical presence and the illusion of authenticity attracted hordes of curiosity seekers from London to Paris, from Marseille to Saint Louis. These indigenous peoples performed daily or ritual activities before the eyes of audiences who could go home at night and sleep in the comfort of their beds after having been around the world not in eighty days, like Jules Verne, but in eighty minutes, as claimed in the promotional slogan for the 1931 Colonial Exposition in Paris.

The distinction between mediated and unmediated presentations of authentic natives is crucial to an understanding of the human zoo phenomenon. Travel narratives report observations on the premise: "I went there and I am now going to tell you about a place you will probably never visit." In some cases, I might even bring back pictures, still or moving. Early photography documented first encounters with non-Westerners and fed a growing curiosity for the remote places and peoples of the planet that would eventually be framed in terms of ethnography and anthropology. In theory, native cultures could be viewed in their authentic environments although, as anthropologists acknowledge, the viewing presence itself alters the native environment. Spectators at ethnographic displays were satisfying their anthropological bent to be observer-participants à la Franz Boas or Margaret Mead. Objects in curiosity cabinets or museums, fashion, stories, music and the like all washed back onto the shores of Europe and North America and provided a glimpse of the larger world. This glimpse was, however, just a glimpse without context. The display stripped the scene of a necessary background and environment against which to measure and understand the foregrounded subjects. Not unlike the lion or the giraffe removed from the African savannah and placed in a metropolitan zoo, perhaps framed by an artist's imagined rendering of the animal's absent setting, indigenous peoples were deterritorialized and inserted into the familiar environs of their conquerors and masters. Their presence made the conquest visible and alive. The return of the exotic object often sent ripples back through the metropolitan culture, as Théophile Gautier commented after the French conquest of Algeria:

> Chose étrange! nous croyons avoir conquis Alger, et c'est Alger qui nous a conquis. — Nos femmes portent déjà des écharpes tramées d'or, bariolées de mille couleurs qui ont servi aux esclaves du harem, nos jeunes gens adoptent le bournous en poil de chameau. (19)

The direct consumption of fetishized objects from the colonial world was then as it is now a mainstay of the fashion, design and art worlds. The multiple sites for representing the Other all have in common that they are mediated representations that viewers and eventually consumers were expected to accept as authentic and truthful without firsthand observation or verification. But if we could actually see the Other close up, would we not believe in their existence, appreciate their difference, better understand who we are because of what they are? The thirst for authentic, moving, flesh-and-blood natives, a "living canvas" of humanity, as it was termed in the catalog to the 1889 Paris World's Fair, set off one of the most important and most successful popular culture movements in modern times: the human zoo. "Seeing is believing" became the motto and the sales pitch for the millions of Euro-Americans who would flock to see Congolese pygmies, Kanak cannibals, Osage Indians, displaced villagers of all sorts, Amazons, Polynesians, Tourags, Eskimos, and myriad other peoples brought to the centers of metropolitan power for the sole purpose of being put on display.

The hey-day of the human zoo ran from the 1870s through the early 1930s when cinema established itself as a worthy rival to reality and fascism quickly came to occupy the attention of the world. Over the course of roughly sixty years, hundreds of these ethnographic spectacles were mounted around Europe, North America and even across the Commonwealth and the colonial capitals. The ticket sales for three of the major Paris exhibits attest to their success: 1889 (32 million), 1900 (48 million), 1931 (33.5 million) (MacKenzie 193). This period corresponded to the rapid expansion of European colonial power around the globe. The 1885 Congress of Berlin led notably to the division of Africa into zones of influence accorded to major European powers. The scramble for Africa was on. The last half of the nineteenth century also coincided with the elaboration of a purported science of race. Count Gobineau's 1855 "Essai sur l'inégalité des races humaines" promoted in pseudo-scientific terms the notion of superior and inferior races based largely on physical criteria and skin color. Colonization led Westerners beyond idle curiosity and speculation about the Other to a phase where foreign lands and peoples had to be dominated through a range of commercial, religious, administrative and military practices. The human zoo phenomenon, placed within the triumphant visual rhetoric of the colonial exhibition, participated significantly in that process.

As the novelty of exotic animals waned, zoo directors sought ways to shore up attendance. Enter Karl Hagenbeck who had achieved fame by furnishing a large number of the wild animals that were exhibited in the metropolitan zoos that were opening all over Europe and North America. His own zoo outside of Hamburg created the prototype common today in many Western zoos where animals are displayed in a man-made natural setting without cages and bars. The very phrase "man-made natural setting" may seem like a contradiction in terms. For Hagenbeck's ground-breaking Tierpark in Hamburg-Stellingen, it meant creating a replica of an animal's natural environment while keeping spectators at a safe distance thanks to moats and other man-made "natural" barriers and enclosures. If "environment" means "interconnectedness" (Morton 7), one might more accurately call these enclosed spaces "décors." While they attempt to replicate facets of the animal's authentic surroundings, their main concern is to stage the animal for the human observer rather than promote the animal's connection with the range of flora and fauna it might encounter in its natural environment. It is a simulacrum, pure and simple. In 1874, Hagenbeck, always the consummate entrepreneur, decided to bring groups of Samoan natives and Laplanders to be exhibited alongside exotic animals at his Stellingen zoo. It was a huge success, so he followed up in 1876 with an exhibit of Nubians from the Sudan and thus discovered a formula for gaining back an audience eager to take in the latest and most sensational live displays. In Hagenbeck's defense, it is worth pointing out that he was very business-like in his dealings with his subjects, drawing up contracts that specified their salary, health benefits, working conditions, etc. Geoffroy de Saint-Hilaire, director of the *Jardin d'acclimatation* in Paris, organized in 1877 his own Nubian exhibit that soon doubled attendance and drew nearly a million paying spectators in one year. The press touted the display as "a band of exotic animals accompanied by no less exotic individuals" (Blanchard et al., *"Les zoos humains…"* 64). Up to the outbreak of World War I, the *Jardin d'acclimatation* organized over thirty such "ethnological exhibits," as they came to be known. Besides the officially sanctioned colonial exhibits and worlds fairs that took place across many European capitals, indigenous peoples were also displayed in circuses, fairs, and music-halls such as the famous Folies-Bergère in Paris where a group of African pygmies were exhibited in 1886.

The main venue for exhibiting the exotic Other quickly became the native villages that were a regular feature of the rapidly proliferating colonial exhibits that took place in major and even minor European cities. Among the most popular were what came to be known as the "Senegalese Village" or the "Negro Village." The 1889 Paris World's Fair commemorating the centennial of the French Revolution drew throngs of visitors who admired not only the recently-erected Eiffel Tower but also the 400 Africans living in reconstructed villages that dotted the Champ de Mars, the large esplanade at the foot of the tower. The 1931 Colonial Exhibit featured over 1500 Africans performing in the shadow of full-scale replicas of Angkor Wat, North African markets, and even Washington's home at Mount Vernon. Interestingly enough, these villages were sometimes constructed of raw materials brought from the natives' homelands in a gesture toward authenticity. They thus created an artificial environment that was presented to viewers as a "real" and faithful replica of the absent original. Among the most popular venues built for the 1931 exhibit was the Vincennes Zoo that remains today on the eastern edge of Paris. The display of wild animals brought to the heart of imperial France was mirrored by the display of indigenous colonial subjects whose juxtaposition further fed the flames of the discourse of savagery, often playing on stereotypes of cannibalism and barbarism. Humans and animals were recontextualized by their colonial captors, transformed into objects taken from an imagined environment, placed in décors and settings that had as much to do with European fantasies or phantasms as they did with reality. Their very presence raised the thorny question: do objects emerge as objects because of their environments? Does a subject become an object as a result of its surroundings? Can subjects claim agency in an environment not of their making?

There were literally hundreds of native villages which appeared during the sixty or so years leading up to the 1931 colonial exhibit. In most cases, they revolved around a reconstructed village where visitors could watch natives engaged in daily or ritual activities such as cooking, crafts, sports or dance. Fascination for the belly-dance as well as native arts and foods permeated the crowds of spectators. These "anachronistic spaces" (244-45), to borrow Anne McClintock's term, created and reinforced hierarchies of difference: civilized over savage, modern over primitive, white over black. Displays of archaic practices created jarring contrasts with the surrounding celebration of technological and economic progress that fed European convictions of cultural and racial superiority. Above all, the native villages objectified European views of the other as irredeemably stuck in an inalterable past time. Images of the primitive and the savage became normalized to the point of becoming stereotypes. Hegel declared in 1830: "The Negro represents natural man in all his wild and untamed nature" (qtd in Pieterse 34). Nature was thus depicted as knowable and conquerable, even at a great distance. Europeans gazed upon the specimens of primitive humanity in atavistic terms: these poor souls could only be relics and reminders of some distant past. According to David Hume:

> I am apt to suspect the Negroes, and in general all the other species of men (for there are four or five different kinds) to be naturally inferior to the whites. There never was a civilized nation of any other complexion than white, nor even any individual eminent in action or speculation. No ingenious manufactures amongst them, no arts, no sciences... Such a uniform and constant difference could not happen, in so many countries and ages, if nature had not made an original distinction betwixt these breeds of men. (qtd in Pieterse 41)

Africans and exoticized others were victims of their nature and of their surroundings which determined their inferiority and their fate. Ethnographic displays commodified natives who entered European consciousness as signs of Europe's own distant past, as markers against which to measure Europe's march of progress to a radiant future.

The Paris colonial exhibition served as propaganda aimed at demonstrating the benefits of the 'civilizing mission' and of the progress brought on by France's colonial dominion over large parts of Africa and Southeast Asia. The spectacle of indigenous peoples provided living proof of their difference: physical traits, customs and manners, technology and science, the arts. Their naked or near-naked bodies conveyed a sexuality that was considered by civilized European standards to be primitive but also seductive. It is no coincidence that the later 1920s up to the time of the 1931 exhibit saw the appearance of the African-American blockbuster star, Josephine Baker, performing her energetic and eroticized dances on the stages of some of Paris' most fashionable venues. Baker enjoyed the greater freedom from racism she came to enjoy in Paris and no doubt the French appreciation of her uninhibited charms came in no small part from the fact that she was black and American, authentically other yet not so radically different as to be unknowable. Jazz and dance performances contributed to the growing negrophilia that was sweeping parts of the French capital at the same time as firm racialized lines were being drawn that separated primitive from civilized, inferior from superior, black from white. This dual tension made displays like human zoos into sites of fear and fascination, reflecting a desire and an anxiety when confronted with the staging of difference. They provided a living showcase of the many racial types that existed and could be placed in a well-ordered scheme by Western science.

In a word, Others became typical, no longer unique and anomalous specimens such as Saartjie Baartman, the "Venus Hottentot," or Joice Heth of the earlier nineteenth century but representative types of a broader category that, as I argued earlier, could stand in for the totality of their kind. The panoptic presentation of difference placed the European spectator at the center of an ever-broadening world that revolved around the observer in a pre-Copernican confirmation of cultural and ontological superiority. These presentations of marginalized peoples confirmed Europe's place at the center of the world. The voyeuristic gaze was uni-directional and power-laden. Unlike the motion pictures that paradoxically froze the spectator in his/her seat while watching the kinetic spectacle on a screen, colonial exhibits allowed a dance of binary opposites that comforted the colonial audience in its belief that it was central to the imperial project and that its might was proven by the spectacle it could afford itself. Western powers had succeeded in transforming natives into objects destined to become commodities in a global circulation of goods and images. The performance of difference played a crucial role in grounding that metonymic relationship that characterized the colonial encounter. Placed alongside the accomplishments of Western progress, indigenous peoples were seen either as doomed to disappear or as potentially useful if they could be assimilated into the universal civilizing mission.

The case of Ota Benga stands out here. A Congolese pygmy brought to the St. Louis World's Fair held in 1904 to commemorate the Louisiana Purchase and that particular brand of American imperialism known as "manifest destiny," Ota was put on display as the only authentic cannibal in North American, a claim supported by the fact that his handlers had filed his front teeth to a point. Following the exhibit, Ota returned to the Congo but then came back to the U.S. with his missionary sponsor. He was transferred to the Bronx Zoo in 1906 where he was for some time placed in the Monkey House alongside apes and other primates, even being labeled for a time "The Missing Link." He was a popular display and posed with visitors. After protests and pressure from humanitarian groups, he was released and Ota Benga found his way to Virginia where he must have felt the full brunt of his alienation and finally took his life in a ritual suicide. While we have few records of the inner drama lived by the natives put on display, Ota Benga's dramatic gesture must stand as a shocking testimonial to the extreme psychological effects felt by many of them. Human displays of alterity were both comforting and unsettling to spectators who were implicitly called upon to take up the white man's burden and to bring civilization to the obscure and backward parts of the world. More than any abstract treatise or pictorial representation could achieve, the physical proximity of spectator to spectacle carried a potent charge that challenged all parties' presuppositions about humanity.

The power of the ethnographic spectacle can be explained by some obvious factors that I have tried to evoke: racial discourses, controversies over evolution, colonial ambitions, domestic tensions, national anxieties, and commercial entrepreneurship. The impact of such spectacles stems in no small degree from their ability to stage moving, kinetic displays of difference to on-lookers who are invited to cast a voyeuristic gaze upon the exotic Other. That gaze reinforces the power differential across the divide between spectator and spectacle. As John Berger has famously pointed out in his seminal essay, "Why Look at Animals?," there is no reciprocity across the moat or the bars that separate spectator from object of observation. The marginalization of animals, like the marginalization of colonial subjects, has led to a blinding of the objects of contemplation such that their gaze is never returned. "Yet nowhere in a zoo can a stranger encounter the look of an animal" (Berger 261).

It is no coincidence that the wave of human zoos ended in the early 1930s when cinema took over the monopoly of moving pictures. One of the first films by the pioneers of cinema, the Lumière Brothers, was a series titled the Ashanti Village shot in 1897 at a colonial exhibit in Lyon. On American shores, Edison captured images of the many worlds fairs at the turn of the century, including the 1904 St. Louis World's Fair where Ota Benga was first put on display. After the last major colonial exhibit in 1931, one of the earliest popular film hits was a 1933 thriller about a certain oversized, sentimental ape named King Kong. Suffice it to say he didn't like being put on display. Our sense of wonder stems from the reality of the display in its multi-faceted, multivalent and seemingly unmediated presence. Like the thrill of seeing and perhaps even touching a wild animal, albeit in a controlled environment, seeing and perhaps interacting with peoples from the far reaches of the earth provides its own thrill as recorded in the many remembrances of spectators who attended the turn-of-the-century fairs and exhibits.

Veit Erlman has examined the "spectatorial lust" on which colonial exhibits hinge, and more precisely the "certainty of representation" (120) that he locates at the center of the colonial spectacle's success. This pact relies on an acceptance of the fundamental distinction between representation and reality. The Senegalese village, Ota Benga, or Saartje Baartman were all viewed by Western spectators in a panoptic flattening of a distant world that viewers are invited to contemplate and accept as a slice of reality. These spectacles created orders of knowledge and confirmed preconceived beliefs through which Europeans established themselves as viewing subjects by the gaze they cast upon objectified natives. These decontextualized fragments of other cultures led viewers to become interpreters of cultural signs, arbiters of difference, participants in a parade of radical alterity. They were called upon to declare the unreal slivers of other humanity as satisfactory representations of reality, a reality they could only imagine and construct from the representations they witnessed.

But the raw power of these human zoos still begs for an explanation. Let me suggest that they derive a good deal of their appeal not only from the fact that they move but also from the fact that they circulate. Indeed, human zoos represent the first large-scale assembly of humankind on a global scale. They are fundamentally popular events for mass culture. As colonialism drew its mantle over great swaths of the inhabitable world, so capitalism engaged in the dizzying circulation of raw materials and manufactured goods. I can do no better in this regard than to refer to Stephen Greenblatt's study of mimetic capital in early modern explorations of the New World (6). Adapting his argument to my purposes, I would sum it up as follows:

- First, that the will and the ability to cross great distances in search of new products and profits create the very conditions of possibility for peoples to circulate.
- Second, that each spectacle maintains the essential reproductive quality of capital in its ability to spawn more of its own kind.
- Finally, to quote Greenblatt: "mimesis, as Marx said of capital, is a social relation of production" (6). He means by this that any production is not just a representation of some social reality but that it also modifies and creates new forms of social reality. It becomes both product and producer.

This quick summary of a fairly complex idea will, I hope, allow us to see the human zoos of yesterday not just as representations of a set of historical and ideological conditions but as pieces of a larger, dynamic system that endures to our day.

Works Cited

Berger, John. "Why Look at Animals?" *About Looking*. New York: Vintage, 1991 (1980). 3-30.

Blanchard, Pascal, Eric Deroo, Gilles Manceron, eds.. *Le Paris noir*. Paris: Éditions Hazan, 2001.

______, Nicolas Bancel, Sandrine Lemaire. "Les Zoos humains: le passage d'un 'racisme scientifique' vers un 'racisme populaire et colonial' en Occident." *Zoos humains. Au temps des exhibitions humaines*. Sous la direction de Nicolas Bancel, Pascal Blanchard, Gilles Boëtsch, Éric Deroo, Sandrine Lemaire. Paris: Éditions La Découverte, 2004. 63-71.

Erlman, Veit. "'Spectatorial Lust': The African Choir in England, 1891-1893." *Africans on Stage: Studies in Ethnological Show Business*. Ed. Bernth Lindfors. Bloomington: Indiana University Press, 1999, 107-34.

Exposition Universelle de 1889. Les Expositions de l'État au Champ de Mars et à l'Esplanade des Invalides. Tome 1. Paris: Imprimerie des Journaux Officiels, Quai Voltaire, 31, 1890.

Gautier, Théophile. *La Presse*. January 6, 1845. *Voyage pittoresque en Algérie*. Ed. Madeleine Cottin. Genève: Droz, 1973): 19.

Girardet, Raoul. *L'Idée coloniale en France de 1871 à 1962*. Paris: La Table ronde, 1972.

Greenblatt, Stephen. *Marvelous Possessions: The Wonder of the New World*. Chicago: University of Chicago Press, 1991.

MacKenzie, John. "Les Expositions impériales en Grande-Bretagne." *Zoos humains. Au temps des exhibitions humaines*. Sous la direction de Nicolas Bancel, Pascal Blanchard, Gilles Boëtsch, Éric Deroo, Sandrine Lemaire. Paris: Éditions La Découverte, 2004, 193-202.

McClintock, Anne. *Imperial Leather*. New York and London: Routledge, 1995.

Morton, Timothy. *The Ecological Thought*. Cambridge and London: Harvard University Press, 2010.

Pieterse, Jan Nederveen. *White on Black. Images of Africa and Blacks in Western Popular Culture*. New Haven and London: Yale University Press, 1992.

Claire Keith

Marist College

Pilgrims in a Toxic Land: Writing the Trenches of the French Great War

This article revisits in an environmental perspective the literary perplexities raised by the unprecedented experience of modern mechanized warfare in WWI. It proposes that the dominantly agrarian economy and cultural traditions of France lent to the body of French war testimonies an ecological sensibility and expressive vocabulary that are distinct from those of their British or German counterparts, as they attempt to document the terminally damaged and poisoned environment of the trenches of the Western Front.

In 1908, a few years before going to war and meeting his death in the trenches of WWI, a young medievalist, Henri Chatelain, had pre-sented for his doctoral thesis a critical edition of the fifteenth-century work *Le Mistère de Saint Quentin*. Commenting on the enumerations that were a favored literary device of the times, and of the work's author in particular, and looking to illustrate his point in a way most likely to entertain his reader, he delighted in quoting the following description of medieval warfare:

> N'y seront pourfendus,
> Patibulés, pourbondis, pourboulis,
> Matés, murdris, martelés, morfondus,
> Boutés, baignés, broquiés, brûlés, bruhis,
> Bien bersandés, bertraudés, forbanis,
> Fronciés, fichés, fourdroiés, fatrouillés,
> Croquiés, hunchiés, courbatés, coustilliés,
> crucifiés, craventés, escrolés
> Esservelés et bien dur castilliés,

Galins galans, Gaulois seront galés. (Sandre 601)

That this would prove a tragically prescient description of exactly what the newly mechanized war of 1914-1918 would do to human bodies and psyches is now amply understood, as are the challenges, both narrative and descriptive, that the experience of mass mechanized death has posed to literature ever since.

What has been less noted from a literary perspective is the extent to which this medieval description applies to the fate of the physical and natural environment in this particular conflict, so much so, in fact, that what happened to the ecology of the theater of war can be rightly seen as the ruling force in the human experience of war in 1914-18.[1] The Great War, it should be noted, was quickly defined and is still remembered today by all but military historians for its Western Front, a continuous system of trenches along the Belgian and French borders, rather than for its broader peripheral theaters or its naval and aerial warfare. In comparison to the ever larger and more destructive conflicts that would follow, WWI remains exemplary in its concentration of the particular conditions that made the soldiers' life and death entirely symbiotic with that of the earth around them: the geographic concentration of hostilities; the near immobility of the main front; the reliance on heavy artillery as the principal mode of warfare; the subsequent necessity to move the troops underground; the duration of the conflict; and the first industrial use of chemical weapons.

Before describing further the ecology of these conditions and its literary implications, it matters here to make the case for a French, rather than generally Western (British, American or German) analysis of the experience of a damaged landscape. Most countries who participated in the conflict have seen notable works emerge from the Great War

[1] Works with a central interest in war and the environment are still few and mostly historical, scientific, or with a broader context of global policy. A reader interested in those could consult among others Richard P. Tucker and Edmund Russell's fine collection of essays *Natural Enemy, Natural Ally: Toward an Environmental History of Warfare*; Charles E. Closmann's *War and the Environment: Military Destruction in the Modern Age*, with the chapter therein by Dorothee Brantz "Environments of Death: Trench Warfare on the Western Front"; or Jurgen Brauer's *War and Nature: The Environmental Consequences of War in a Globalized World*.

from the gifted pens of exceptional writer-soldiers — Dos Passos, Hemingway, Jünger, Maria Remarque, Sassoon, Graves or Blunden — and France itself finds among its combatants some of its great literary names: Apollinaire, Duhamel, Genevoix, Giono, Céline. However it is in agrarian France, more than in the rest of Europe, that appeared in the ten years following the war a vast testimonial effort facilitated by the particular historical, geographic, economic and social conditions of that country. In 1983, the popular historian Pierre Miquel still chose to say without hesitation and without being contradicted that "la Grande Guerre reste pour nous par excellence une guerre française"(9), thus reminding us not only of the staggering statistics of its losses (almost 2 millions dead, 5 millions wounded of a population of 39 millions) but also of the fact that unlike the British or German soldier, the French soldier was fighting on his own soil to defend his own homeland.[2] Three decades later and with a founding commitment to comparative international research, the highly visible memorializing and pedagogic enterprise in progress at the Historial de la Grande Guerre in Péronne, in the Somme region, still functions actively on the premises that France's Great War was distinctive within Europe in the national *consent* that fueled its long war effort.[3] In the postwar period in France, then, the memorializing of war was certainly more welcome in France than in England or in Germany where the losses incurred from having intervened in or initiated the conflict could not be assuaged by a physical and emotional link to the land that had been protected. During the conflict and for ten years after, hundreds of memoirs, journals, correspondences and assorted documents were able to get immediately in print in France through the help of diligent veteran associations and with the temporary indulgence of the literary and publishing establish-

[2] Belgium, occupied by the German Army as a passageway to France through most of the war despite its neutral status, experienced and memorialized its war nationally mainly for its painful civilian toll that generated international outrage and helped fuel Allied war propaganda. Belgian combatant testimonies are largely unpublished and only recently analyzed (Amez).

[3] For a quick view of this dominant current in French WWI historiography, the resources on the Historial's main website <http://www.historial.org> are quite consistent in their emphasis on national identity and the concept of *culture de guerre* expounded by its founders (Audoin-Rouzeau, Becker). Alternate scholarly views by French historians tend to mitigate rather than reject the concept outright (Offenstadt).

ment. Several anthologies followed, notably that of one truth-obsessed veteran, Jean Norton-Cru, whose exhaustive and meticulous cataloguing is now the standard bibliographic and critical reference for all reflections on French war testimonies (Norton Cru, *Témoins*).[4]

What this gives us is a raw corpus of texts — la *Matière du Front,* as it were, in a distinctively French manifestation: that of a common literary sensibility and common references for *la belle écriture* acquired by all, peasants and aristocrats alike, in the schools of the Third Republic, and of a deep cultural attachment to the home soil that found its easily recognizable expression through the earth-celebrating language of Joachim Du Bellay, George Sand or Charles Péguy or in paintings such as Millet's *L'Angélus* or Rosa Bonheur's *Le Labourage nivernais*; a fertile, fundamentally agricultural landscape, at peace with God and the elements, and nurtured by a devoted commitment to the ennobling and mutual benefits of mankind's respectful working of the land.

Furthermore, this French testimonial literature, in contrast to that of poetry-loving Britain for example, chose overwhelmingly to exorcise its traumas through the narrative prose of memoirs, autobiography, journals, novels and correspondence, as the most natural format in which to apply the standards of writing extolled universally in French public schools. By the accident of their birth, French war writers had to express themselves in a language that reveres its own tradition of clarity, dignity and above all, of Cartesian precision. What is moving in these French testimonies, and doubly so in an ecological reading given the intense physical bond with the earth generated by these soldiers' circumstances, is the determination with which they strive to maintain in an impossible adversity the tradition of elegant writing, of a broad vocabulary, of the polished phrase, of the best choice of tense, of descriptions that emulate the splendid *dictées* of the *Certificat d'Etudes,* of the magic of the formal *vous*: ancient habits of those for whom a certain elegant formality of the language remains an island of peace amidst the upheaval of the mind. It is not an exaggeration to state that this striving, this reflexive emulation of time-honored writing standards, is felt in all these testimonies and correspondences regardless of education level. The challenge of describing the bizarre natural side effects of a modern

[4] A persuasive defense of Norton Cru's stern critical approach to WWI testimonies can be found in a 2007 piece by the scholar of narrative discourse Charlotte Lacoste.

mechanized war was acknowledged even by the uncontested master of traditional description, Pierre Loti, who in an otherwise one-dimensional period piece of anti-German propaganda from 1916 observed that

> notre chère langue française, qui s'est formée dans la beauté, n'avait pas su prévoir les mots dont on pourrait avoir besoin un jour, au vingtième siècle, pour désigner certaines abominations et certains monstres. (5)

From the paternalistic *imprimatur* of the literary mandarins of the day[5] to the general willingness of the troops to trust their educated officers with the various trench publications representing their lives,[6] it is clear that the use of a proper and even literary language seemed natural and desirable to all to express in official print their perceptions of their odd environment, unlike the popular talk and *argot* that were the norm in the everyday trench. There were still such things as Sunday clothes in 1914 society, and the letters from the least educated soldiers were the ones most often framed in the highly codified formulas borrowed from the administrative correspondence that punctuated a humble rural life: birth certificates, census, update of military papers. Having submitted with docility to such rituals of the Republic as the *Procès verbal d'échange du fascicule de mobilisation*: "Cejourd'hui, nous, gendarme à .... , avons inséré dans le livret individuel du dénommé ci-dessus, en échange du présent fascicule ... " (*Manuel* 846), a peasant-soldier would spontaneously sign a letter to his parents: MARTEAU, Emile.

This legacy of precise formality and of abundant and familiar literary metaphors for a domesticated nature would prove both a help and a handicap when the time came to report, for instance, the terminal churning of clay, chemicals, water, metal, blood and bones that was to be the final and only proven outcome of the eight month battle of Verdun. Before addressing this point further, it will be useful here to raise the

---

[5] "Comment ceux d'entre eux qui se sentaient un don d'écrire, même très faible, résisteraient-ils à la tentation de les noter? Certes, ces livres sont d'une valeur bien inégale. Ne regrettons pas leur abondance." (Bourget, *in* Jubert,14.)

[6] On the socio-cultural background of most editors of newspapers from the trenches, see Audoin-Rouzeau, *14-18, Les Combattants des tranchées*, 14.

issue of self-censorship, it being such a default reflex in a fighting soldier anxious to avoid official trouble but also unnecessary anguish to their loved ones. With the war's dull horrors increasingly in need of corrective propaganda, letter-writers in the trenches as well as wartime reporters and memoirists had to tread carefully when reporting on sensitive topics, the long list of which included death and modes of dying, the rate of casualties, authority, heroism of the real or manufactured kind, the relative notion of cowardice, frontline lucidity about the ratio of bellicosity to personal risk, and a general disillusion in the face of war among soldiers of any political persuasion. On the other hand, there seems to have been little perceived need to impose a gag order on the telling of environmental destruction. For that matter, a quick survey of painting, this other witness art of the war, shows similar conditions surrounding officially commissioned work, private sketchbooks and the immediate post-war production, all of which reflect various levels of intent and freedom to reveal or minimize the graphic evidence of *human* death and suffering. Censorship of the testimonial paintings mirrors that of the written testimonies, as artists were criticized variously for showing too much or too little of the human misery of war, whereas no serious controversy was attached to the chosen interpretations of landscape except relative to the perspective just stated.[7] Both officially commissioned war artists and independent or dissenting artists could depict more or less as they saw fit the concentrated war garbage littering villages, the endless bombardment ruins and the oddly altered appearance and topography of the rural landscape, as these were not received *per se* as an indictment of war.[8] On the whole, then, official censorship as well as family-soothing self-

[7] For a general introduction to war painting with a French perspective, see the article by Marie Pascale Prévost-Bault, Curator in Chief at the Historial de la Grande Guerre; the site of the Mémorial de Caen <http://www.memorial-caen.fr/10EVENT/EXPO1418/fr/peintre.html>; and the deeply moving site devoted to soldiers sketchbooks <http://www.dessins1418.fr>. For an overview of the larger research field of documentary war photography, also subject to censorship, see <http://www.guerre1418.fr/grande-guerre-14-18-photos>.

[8] For the pictorial use of environmental destruction as a dedicated and powerful indictment of war, the contemporary artist Jacques Tardi has established the canonical reference of our time with his treatment in graphic novel format of the trenches of the Great War, positioning his work as the graphic counterpart across the decades of Henri Barbusse's classic *Le feu*.

censorship was mostly vigilant about the human side of war — the legitimacy of it and the realistic depiction of the sufferings it inflicted on people. One needs only observe the place of horses in the general economy of the Great War to remember that crimes against nature and animals had not yet been conceptualized in legally criminal terms and that the death of nature, unlike the death of men, could be deplored collectively and communally without too much political risk. Perceived as it still was as an inevitable corollary to war, its literary or visual representation could even appear to officials as a properly solemn evidence of the seriousness of the conflict and as a public reminder of the abnegation of the troops, not to mention the barbarity of the enemy.

Without the pressure of a legally sanctioned ecological morality, but with the more visceral moral response of French men with a deep-seated agrarian sensibility, trench soldiers could and did apply themselves with some measure of spontaneity to making literary sense of the relentless transgressions against the natural order in which they were the helpless participants.

What sort of experimental calamities, then, visited the earth in the environment of the Great War to so perplex the literary witnesses who tried to make descriptive sense of them? In short summary: from an event that was supposed to be a quick, above-ground human affair, the war became instead a monstrous struggle of attrition between the Machine, as understood as the full power of modern industrial artillery, and the physical environment, as the principal recipient of this applied enterprise of destruction. The ratio of human death to ecological death (it took tons and tons of fairly random shelling to decimate one platoon in a trench) makes this arguably one of the most inefficient wars in the history of mankind, creating at every level, be it political, military or civilian, a unanimous sentiment of consternated helplessness. "I don't know what is to be done. This isn't war," complained Lord Kitchener, the British Minister of War. What he was referring to was the deadlock of the Western Front, a vast open wound in the earth that stretched from the Belgian coast to the Swiss border and where two armies, after the failure of the first offensive, had had to entrench themselves a few hundreds of yards apart, giving birth to the troglodyte world first described in a hyper-realistic scatology of death and destruction by Henri Barbusse in his literary grenade of 1916, *Le Feu.*

> La terre! Le désert commence à apparaître, immense et plein d'eau, sous la longue désolation de l'aube [...] tout est mouillé, suintant, lavé, naufragé, a l'air de couler. On distingue de longs fossés en lacis où le résidu de la nuit s'accumule. C'est la tranchée. Le fonds en est tapissé d'une couche visqueuse d'où le pied se décolle à chaque pas avec bruit, qui sent mauvais autour de chaque abri à cause de l'urine de la nuit. Les trous eux-mêmes, si on s'y penche en passant, puent aussi, comme des bouches. (11)

For four years the front lines remained virtually unchanged, moving back and forth by a few hundred yards during murderous and futile attacks. Occupied by the German on one side and by the French and British on the other, the trench system was separated by the infamous No man's land, symbol of the exile of mankind from its formerly above-ground environment, and whose enduring power on the imagination eventually turned it into a generic term of discourse. Scientific atrocities, in the form of immense guns and the new flame throwers and chemical weapons, were beginning their career. There was no question of strategy or tactics. Instead, as Pierre Miquel summarizes, the only remaining option was to eliminate the enemy as one gets rid of vermin: you smoke them out, you asphyxiate them, you burn them, you mine them, you drown them (184). One of the constants of the war of attrition was a prodigious outpouring of shells, month after month, on the same terrain, soon poisoning the soil to a terminal saturation of human remains, metal, excrements, chemicals and the daily detritus of life, so that the ecology of the trenches mostly struck the mind and senses dumb in its singular impropriety for sustaining life, to the exception of a thriving population of rats, the constantly reported nightmare of the frontline soldier. One writer, Pierre Chaine, decided altogether to write his memoirs from the perspective of one, offering in *Mémoires d'un rat* a bio-practical perspective of the front as a new and endlessly replenished feeding ground, thus softening through comic relief an all too evident and revolting aspect of the trench environment. In a similar and most deliberate effort to address the need for new terms and concepts, Jules Romains, the great humanist writer, notes how the war of attrition was sustained by a human paste, a new, malleable and renewable biological substance not unlike wood paste, conveniently dispensed in a continuum, to be consumed and processed like a natural resource, used with a partitive article and called "du million d'hommes":

> Ce front ennemi devant vous. Du million d'hommes étalé en couche épaisse. Il ne suffit plus de le percer au bon endroit. Il faut l'user tout du long. Il faut enlever ça comme à la lime, comme à la râpe, comme à la varlope. Du copeau de million d'hommes (11).

The very limited theater of war made even more evident the strangeness of its ecology in relation to the normal landscape only a few kilometers away. The universe of the front was closed, self-contained, intensely absorbed in a single preoccupation that consumed most of the physical and mental resources of the individual: the frightful interdependency of human death and environmental death in ways that plainly exceeded, in its everyday happenings, anything that centuries of fantastic tales or the imaginations of Bosch and Dante had so far conjured up. Since the imaginative dread of the trenches is often attached to the first large scale use of chemical warfare, a few remarks are in order to place this aspect of ecological destruction in its contemporary context. The Great War was certainly a "chemist's war" in its legally unencumbered experimentation with chemicals, but gas attacks, despite their initial terrorizing power and visually impressive offspring of colored clouds and hideous masks, were in the end less lethal to both humans and nature than the sheer volume of conventional munitions, as the former dissipated quickly and were soon adequately protected against whereas the latter so saturated the land that they could not be avoided and are still oozing out of the land today, mostly as unexploded ordnance.[9] War writings reflect this fact in their overwhelming focus on the

[9] The long-term impact of WWI chemical and conventional ammunitions on bio-health remains very serious: the "Zone rouge" of 11 departments of the war's Western Front counts areas with the highest concentration of heavy metals in the world and lead a ubiquitous presence in the soil. To the dismay of a few scientific groups, the problem remains to this day largely unaddressed in public conversation and policies and continues to be neglected by historians, unlike all other sequels of war — psychological, economic, political, cultural — that are extensively analyzed.

relationship of the soldier with the basic elements: earth, water, air and light; and with the sensory rearrangement of their passive, bombarded, troglodyte life, in which sight cedes place to hearing as the essential tool of survival and smell must learn a new taxonomy of odors to bring cognitive order to its environment.

A comprehensive environmental reading of France's testimonial literature of the Great War will require a patient and cumulative examination of texts that have been mostly read until now to search for the expressive challenges presented to literature by the new scale of mass human tragedy and death. In contrast, such an environmental reading will need to clarify persuasively where and how literary expression responds to distinct ecological stimuli rather than to the general cognitive stress inflicted by war and death. The shortness of the present study does not permit to conduct such a methodical exposition of so broad a material, but some suggested lines of inquiry will be offered. An ecocritical reading will usefully note as a starting point the persistence to this day of national sensibilities in the analysis, representation and memorializing of the Great War, as they will inform the appreciation of a distinct French experience of the destruction of nature in contemporary original war texts. These national differences we take to be most luminously illustrated by a revisiting of the two classic Anglo-Saxon studies (Leed, Fussell) that established in an inspired, impressionistic and essential approach the cultural foundation of most if not all WWI research in English since the 70s;[10] and by a contrasting observation of the didactic, data and object-driven, and more conventionally memorializing effort that underlies the scholarly supervision of the Péronne Historial in France since its creation in 1992 as a committed international project.

Our ecocritical study will want to include the following elements, which we insist must be mindful of the fact that the French soldier, unlike the others, was fighting on his own homeland: a revisiting of the traditional initiatory entry into war from the critical perspective of the landscape and the natural world; an attentive analysis of the language

---

[10] For a thoughtful salute to these two seminal works and their influence, see the entry in the blog *WWI Bridges* posted shortly after Paul Fussell's passing in 2012 <http://wwiindex.blogspot.com/2012/05/great-war-and-modern-memory-some.html>.

chosen for sensory reports; an inventory of the metaphors, old and new, used in trying to apply trusted writing practices learned in school to the war's odd bio-scenery; a comparative reflection on what is universal and what may be particular in the literary and cultural processing of ecological disruption, including the suitability of prose over other forms to this end and the response to climate and topography by writers of different regional and cultural backgrounds; a recognition of the dominant role of earth and water in all these narratives.

As an introduction to the literary gems still awaiting full attention in a nature-oriented study, we will propose a few passages gleaned from some of the best known and from a few lesser known works of the *écrivains-combattants.*

Like men, the landscape was still intact at the date of August the 22nd 1914, and Jean Galtier Boissière's *La Fleur au fusil* (1917) is among the few narratives to dwell at some precise length on the startling moment of transformation of the civilian landscape into a war landscape.

> Des explosions d'une violence inouïe déchirent l'atmosphère et se répercutent longuement sur les hautes futaies. Des branches brisées fouettent le sol de leur feuillage frissonnant ; des arbres oscillent, fendus en deux comme par une hache invisible; fauchés comme des fétus, les géants de la forêt s'écroulent avec des craquements effroyables, écrasant dans leur chute tout ce qui vivait sous leur ombre (105).

A few months later and for years to come, there would be not one tree left on the entire line of the Western Front. In the classic *Voyage au bout de la nuit*, Louis-Ferdinand Céline also invests all his hatred of war in this first moment of discovery and expresses it similarly through the betrayal of a familiar landscape, this time making nature complicit:

> Moi d'abord, la campagne, faut que je le dise tout de suite, j'ai jamais pu la sentir, je l'ai toujours trouvée triste avec ses bourbiers qui n'en finissent pas, ses maisons où les gens n'y sont jamais et ses chemins qui ne vont nulle part. Mais quand on y ajoute la guerre en plus, c'est à pas y tenir. Le vent s'était levé brutal, de chaque coté des talus, les peupliers mêlaient leurs

> rafales de feuilles aux petits bruits secs qui venaient de là-bas sur nous. [...] De temps en temps, je ne savais d'où, une balle, comme ça, à travers le soleil et l'air, me cherchait, guillerette, entêtée à me tuer, dans cette solitude, moi. Pourquoi? Jamais plus, même si je vivais encore cent ans, je ne me promènerai à la campagne. (13)

In his novel *La Percée*, Jean Bernier explores the same discovery in an entirely different prose, highly intellectualized, almost precious, and profoundly convincing. Favigny the soldier arrives at the front at the moment when the propaganda of the rear is still in full bloom but when the reality of war reveals itself ominously through a sticky, clayey antechamber of Death, "cette boue isabelle qui sourdait sous les pas, juteuse, visqueuse, incompressible" (28) and which, like war, devours everything. Glimpsing some survivors of the first August campaign, broken, colorless, he suddenly wonders in fear how he will tell this, he who had planned to write about a heroic war:

> Comment leur dire cette cachexie universelle, cette dégradation putride des routes, des murs ou des étoffes, des hommes; le déluge de ces formes, vivantes ou mortes, comme plongées déjà dans quelque gras cimetière? (30)

All testimonies give ample evidence that whatever eventual adaptation could be made to trench life was mostly of an environmental nature. Since nothing could be done to alter the course of shelling, resourcefulness defined itself in the form of an intimate accommodation with the earth. This is described as an instinctive but often futile attempt to hide under it against an arriving shell, or in a much more profound and final way, as an acceptance expressed less in a resigned vision of one's brutal dismemberment as in a full and willing surrender to becoming one with the putrid earth. As a modest contribution to the roaring symphony of the shelling, Jean Bernier composes for the lonely soldier a little lullaby that he names *Chanson du tir de barrage*. Line after line, the bittersweet litany exhorts the soldier to dig always faster and feverishly that" hole that makes him one with the clay and where he will await his fate: "Aime la terre ... bénis … à plat ventre … travaille … Rejette …" (Bernier 199)

In the tellingly titled *La Boue des Flandres*, Max Deauville poignantly recasts his earlier perception of nature as a vain mirage:

> Nous ne voyions jadis la terre que couverte de son manteau magnifique. Ses campagnes vertes épinglées de maisonnettes roses, ses eaux vives, ses rivières entourées de buissons n'étaient qu'un délicieux mirage [...] La nature pour nous se concrétisait dans la parure des prés... (315)

As the conflict progresses, earth, like men, succumbs to the violence that dislocates it, kills it, and leaves it to rot indecently in the view of everyone, to be finally transformed into a dead dust. All trench witnesses exhibit an extremely crisp consciousness of this ugly agony and they document it with care, often using a terminology of rape and religious sacrilege: "La glèbe naguère fertile, souillée maintenant de poisons, sa chair morte, inguérissable de notre immonde supplice" (Genevoix 514). Or as René Naegelen writes: "Une odeur de chimie et de charogne monte de la plaine mouillée et les obus pleurent dans l'air la détresse poignante de la nature assassinée" (165). Earth itself changes form under the leveling power of shells, hills lose their contour, their very mounds:

> A Verdun, une division, dans l'espace d'une relève, laisse en moyenne 4000 hommes. La terre elle-même change de forme: les collines sous les coups de rabot des obus, perdent leur relief, leurs contours. Le paysage prend cet aspect monstrueux, jamais vu, cet aspect de néant, cette apparence croulante de fourmilière et de sciure où des échardes, des fétus, des débris de choses mêlées comme de la paille dans du mauvais pain, rappellent qu'il y a eu des bois, des fusils, des brancards, on ne sait quoi de concassé là. (Gillet 49)

Anyone who discovered suddenly and without warning such transformations probably shared the sentiment of total stupor expressed by a an auxiliary medic who arrived shortly after the mine explosion of Vauquois, remembered as one of the most powerful of the entire war, that resulted in an entire hilltop removal:

> Je fus à cet autre point de vue, d'où on me dit que se voyait Vauquois. Je regardais. Je fus hébété. Je savais qu'il ne restait rien de Vauquois. Je ne m'attendais pas à ce spectacle. Dominant tous les environs, une croupe lisse et blanche, un crâne, un oeuf. Si quelque chose peut donner une impression pareille de dévastation, de suppression, je voudrais savoir où (Langerhager 67).

The unfamiliar landscape clearly leaves the author at a loss for references, as the stock of similes collected through his literary schooling to describe mountains is of no help. He has to make no less than three attempts whose prosaic simplicity would no doubt make his old schoolteachers frown, who were entitled to expect better. To us, of course, they are telling of the emergence of a new consciousness, adapting to a changing ecology whose parameters are not just figuratively strange but scientifically so.

Water, the most cruel and powerful enemy of the trench soldier and yet his most precious commodity eventually proves the winner of even the almighty shelling Machine after days of relentless rains, leaving this vision of silent apocalypse offered by Henri Barbusse: "On voit des lacs, et entre ces lacs de lignes d'eau laiteuses et stagnantes. L'eau a tout pris [...] l'inondation est universelle. Le champ de bataille ne dort pas. Il est mort." André Pézard, perhaps the most piercingly memorable of all the French tellers of the Great War, a felicitous observer of colors, smells, sensations and of the desolate and universal degradation of man and nature in war, must be given here the task to close this brief literary glimpse of the heart-breaking, murdered environment of the Western Front. He who keenly observed "ce chaos toujours renaissant, toujours plus lamentable, toujours plus noir" (201), who revealed the infinite disorder of the war paraphernalia of rusted metal, tangled barbed wire, broken planks and sticks, bent rails, shredded rags with these startlingly evocative words: "On se dirait dans un cauchemar sec et compliqué, dans des cercles de calculs abstrus, dans des combinaisons arides de lignes et de chiffres qui fourmillent sous la clarté plate de l'abstraction" (131), he also finds the final words to express on behalf of all his war companions the desolate affection and memories that a four-year intimacy with loss in a destroyed earth has left in the heart of those who returned. A soft image of spent fruits falling from their tree closes his book *Nous autres à Vauquoix*:

> Laissez-moi dire ceci lentement, comme est lente une pensée endolorie; laissez-moi dire lentement, comme tombent à regret, de chère syllabes meurtries: «Adieu, ma pauvre guerre!». Et c'est tout.
>
> - *Adieu, ma pauvre guerre.* (348)

Works Cited

Amez, Benoît. *Dans les tranchées. Les écrits non publiés des combattants belges de la Première Guerre mondiale. Analyse de leurs expériences de guerre et des facteurs de résistance.* Paris: Publibook, 2009.

Audoin-Rouzeau, Stéphane. *14-18. Les Combattants des tranchées*. Paris: Armand Colin, 1986.

______, and Annette Becker. "Violence et consentement: la 'culture de guerre' du premier conflit mondial." *Pour une histoire culturelle*. Ed. Jean-Pierre Rioux and Jean-François Sirinelli. Paris: Seuil, 1997. 251-71.

Bernier, Jean. La *Percée.* Paris: Albin Michel, 1920.

Brauer, Jurgen. *War and Nature: The Environmental Consequences of War in a Globalized World.* Lanham, MD: AltaMira Press, 2009.

Céline, Louis-Ferdinand. *Voyage au bout de la nuit*. *Romans.* T.II. Paris: Gallimard, 1977.

Chaine, Pierre. *Mémoires d'un rat.* Paris: L'Œuvre, 1917.

Chatelain, Henri, ed. *Le Mistère de Saint Quentin.* Saint Quentin, 1909.

Closmann, Charles E., ed. *War and the Environment: Military Destruction in the Modern Age.* College Station: Texas A&M University Press, 2009.

Deauville, Max. *La Boue des Flandres.* Bruxelles: Maurice Lamertin, 1922.

Duhamel, Georges. *Civilisation.* Paris: Mercure de France, 1918.

______. *Vie des Martyrs.* Paris: Mercure de France, 1917.

Fussell, Paul. *The Great War and Modern Memory*. London, Oxford & New York: Oxford University Press, 1975.

Galtier Boissière, Jean. *En rase campagne. Un hiver* à *Souchez.* Paris: Berger Levrault, 1917.

______. *La Fleur au fusil.* Paris: Editions Baudinière, 1928.

Genevoix, Maurice. *Ceux de 14.* Paris: Flammarion, 1950. Comprising the five following works, originally published at the dates indicated: *Sous Verdun.* Paris: Hachette, 1916. *Nuits* de *guerre.* Paris: Flammarion, 1917. *La Boue.* Paris: Flammarion, 1921. *Les Éparges.* Paris: Flammarion, 1923.

Gillet, Louis. *Un type d'officier français: Louis de Clermont-Tonnerre, commandant de zouaves (1877-1918).* Paris: Perrin, 1919.

Hoffmann, Frederick J. *The Mortal No: Death and the Modern Imagination.* Princeton: Princeton University Press, 1964.

Jubert, Raymond. *Verdun.* Paris: Payot, 1918.

Lacoste, Charlotte. "L'Invention d'un genre littéraire: *Témoins* de Jean Norton Cru." Web. *Texto!* vol. XII, n°3 (2007). <http://www.revue-texto.net/index.php?id=54>

Langenhager, Jean de. *Un soldat* de *France. Lettres d'un médecin auxiliaire.* Paris: Plon, 1919.

Leeds, Eric J.. *No Man's Land: Combat and Identity in World War I.* Cambridge: Cambridge University Press, 1979.

Loti, Pierre. *La Hyène enragée.* Paris: Calman Lévy, 1916.

Manuel d'Infanterie à l'usage des sous-officiers, des candidats sous-officiers, des caporaux et élèves caporaux. Paris & Limoges : Lavauzelle imprimeur militaire, 1914.

Miquel, Pierre. *La grande guerre.* Paris: Fayard, 1983.

Naegelen, René. *Les Suppliciés.* Paris: Éditions Baudinière, 1927.

Norton Cru, Jean. *Témoins.* Paris: Les Étincelles, 1929.

Offenstadt, Nicolas et al. "À propos d'une notion récente: la 'culture de guerre.'" *Guerres, paix et sociétés, 1911-1946.* Neuilly: Atlande, 2004. 667-74. <http://crid1418.org/espace_scientifique/textes/culture_de_guerre.htm>

Pézard, André. *Nous autres* à *Vauquois.* Paris: La Renaissance du Livre, 1918.

Prévost-Bault, Marie-Pascale. "La Peinture et la Grande Guerre." *Les chemins de la mémoire* vol. 136, 2004. <http://www.chemins dememoire.gouv.fr/fr/la-peinture-et-la-grande-guerre>

Romains, Jules. "Prélude à Verdun." *Les Hommes de bonne volonté.* Paris: Flammarion, 1938.

Sandre, Thierry, ed. *Anthologie des écrivains morts* à *la guerre, 1914-1918.* Amiens: Éditions Malfère, 1924-26.

Tardi, Jacques. *Le Trou d'obus.* Epinal: Imagerie Pellerin,1984.

Tucker, Richard P., and Edmund Russell, eds. *Natural Enemy, Natural Ally: Toward an Environmental History of Warfare.* Corvallis: Oregon State University Press, 2004.

Marie Chantale Mofin Noussi

Linfield College

Café toxique: éco-colonialisme et monoculture dans *La Terre du café* de Patrice Nganang

Cet article analyse l'éco-colonialisme, encore appelé colonisation écologique dans la nouvelle *La Terre du café. Lettre à mon grand-père à propos de la France qui ment* de Patrice Nganang. La politique coloniale française d'assimilation s'est étendue à la sphère écologique à travers la mise en place de la monoculture du café. L'introduction des plantations coloniales a contribué à modifier les structures socio-écologiques des colonies et à aliéner les colonisés de leur propre milieu. Nganang utilise une écriture écologique, privilégiant les relations entre différents éléments pour subvertir l'homogénéisation de la politique coloniale.

La littérature écrite d'Afrique sub-saharienne prend son envol dans le conflit de la rencontre avec l'Occident. Cette littérature s'est consacrée en grande majorité aux questions liées à la libération culturelle et politique des peuples du continent africain. Ainsi, les écrivains se sont donné pour tâche de valoriser les cultures africaines et de dénoncer les pratiques colonialistes qui visaient à l'assimilation culturelle de populations africaines. Cependant, il y a un domaine de la vie des colonisés qui a autant souffert de la domination coloniale que les cultures africaines: il s'agit de la terre, et notamment l'agriculture.

La colonisation de la terre est très souvent évoquée par les écrivains comme manifestation de la domination occidentale dans les colonies. Conquérir le plus grand nombre de terres possible était un but majeur des colonisateurs car la possession des terres est une base indéniable pour l'assise du pouvoir. Posséder la terre, signifie être maitre des lieux et avoir le pouvoir sur tout ce qui s'y trouve. Malheureusement, cette colonisation "foncière" est parfois négligée par la critique qui se consacre plus aux questions plus directement liées aux cultures. Karla Armbruster et Kathleen Wallace remarquent d'ailleurs la primauté dans la critique de la culture sur la nature: "culture is the main purview of literary studies, while nature is, if present at all, merely a backdrop of human drama" (4). Notons que la question de la colonisation de la terre est étroitement liée aux cultures dans la mesure où l'environnement et les cultures humaines s'influencent mutuellement comme l'expliquent Armbruster et Wallace. Pour elles, il est important de comprendre "nature and culture as interwoven rather than separate sides of a dualistic construct" (4).

Le texte de l'historien Alfred Crosby intitulé *Ecological Imperialism: The Biological Expansion of Europe, 900-1900*, est sûrement l'un des pionniers dans la mise en évidence de l'aspect écologique de l'impérialisme européen. Parlant de Crosby, Edward Said affirme:

> Crosby says that wherever they went, Europeans immediately began to change the local habitat; [...] this process was never-ending, a huge number of plants, animals, and crops as well as building methods gradually turned the colony into a new place, complete with new diseases, environmental imbalances and traumatic dislocations for the overpowered natives. A changed ecology also introduced a changed political system. (225)

Crosby révèle que l'objectif de l'impérialisme ne se limitait pas uniquement à la "mission civilisatrice" tel que les conquérants voulaient le faire croire. Il s'agit plutôt, selon Crosby de transformer les territoires nouvellement conquis en introduisant de nouveaux éléments naturels tels les plantes, les animaux et même les maladies.[1] Même si l'impérialisme est initialement un système politique, il se sert largement de la

[1] Dans *The Heart of Redness* du Sud-Africain Zakes Mda, le narrateur explique que les colonisateurs sont arrivés en Afrique du Sud avec des virus destinés à affaiblir les populations afin de mieux les dominer: "White people knew of lungsickness because it came from their country. There were reports that it had killed many cattle across the seas in the land of the whites. It was brought to the land of the amaXhosa nation by Friesland bulls that came in a Dutch ship two years earlier, in 1853. Therefore, even the best of the isiXhosa doctors did not know how to cure lungsickness" (50).

géographie, et de la terre dans le cas de cette étude pour installer et conserver l'hégémonie de l'impérialiste. En définissant l'espace géographique en relation avec la colonisation, Eric Wagner déclare que le paysage (compris comme espace géographique) est "the physical place on which the conflict between the colonizers and the colonized is enacted" (573). La préoccupation de ces auteurs est de mettre en exergue la connexion forte, quoique parfois incomprise entre l'impérialisme ou la colonisation et l'environnement; afin de montrer comment l'environnement est utilisé pour subjuguer les populations locales.

Dans la nouvelle intitulée *La Terre du café. Lettre à mon grand-père à propos de la France qui ment* tirée du recueil *Dernières nouvelles du colonialisme*, le Camerounais Patrice Nganang met en dialogue la critique de la situation coloniale et l'état de violences écologiques qu'ont subi les terres de l'ouest du Cameroun et leurs habitants. *La Terre du café* c'est une lettre d'un petit-fils habitant en Occident à son grand-père décédé. En évoquant ses souvenirs de son grand-père, le petit-fils narrateur revient sur les méfaits de la colonisation dans l'ouest du Cameroun qui ont donné lieu à un important bouleversement socio-écologique dans la région. La nouvelle identifie la monoculture du café comme étant le maillon central de la chaîne de destruction coloniale.

Frieda Knobloch dans *The Culture of Wilderness*, stipule clairement au début de son essai que "colonization is an agricultural act. It is also an agricultural idea" (1). En considérant la colonisation sous cet angle agraire, les questions suivantes surgissent dans le cadre de l'analyse de l'œuvre de Nganang: quels sont les impacts de la colonisation française dans l'ouest du Cameroun? Comment s'est manifesté le processus d'assimilation écologique au Cameroun? Comment le style de Patrice Nganang contrecarre-t-il le système d'homogénéisation de l'éco-colonialisme français? Cette étude est fondée sur la thèse selon laquelle le traitement ou mauvais traitement de l'environnement par les colonisateurs français à l'ouest du Cameroun était étroitement lié à la politique française dans les colonies. En d'autres termes, le projet colonial concernant la monoculture est un corollaire de la politique d'uniformisation autrement nommée assimilation. L'assimilation dans les colonies françaises est très souvent abordée sur le plan culturel et linguistique en particulier pour montrer la volonté des colonisateurs d'imposer le français comme langue unique dans les territoires conquis. L'assimilation écologique qu'engendre la monoculture peut facilement être sousestimée, ceci en grande partie parce que les liens entre l'écologique et le culturel ne sont pas parfois perceptibles au premier regard. Pourtant, la monoculture est capable à elle seule de perturber tout le système politique, économique, social et culturel d'une région. L'éco-colonialisme, par l'introduction de la monoculture dans les colonies, sonne le glas de la diversité sur les territoires colonisés. Cette étude examine les manifestations et les conséquences de l'éco-colonialisme dans *La Terre du café*, ainsi que l'écriture écologique utilisée par Nganang pour résister à la destruction de la biodiversité.

Il ne serait pas erroné d'affirmer que l'assimilation est une monoculture, car la politique française d'assimilation fonctionne selon un mécanisme identique à celui de la monoculture. En effet, les deux systèmes ou machines attribuent leur succès à leur capacité de phagocyter toutes les autres cultures (entendues comme mode de vie ou culture sur le plan agraire) pour en faire triompher une seule. On se retrouve ainsi dans le règne de l'anti-altérité décriée par Claude Lévi-Strauss dans *Tristes Tropiques*:

> [L]a civilisation n'est plus cette fleur fragile qu'on préservait, qu'on développait à grand-peine dans quelques coins abrités d'un terroir riche en espèces rustiques, menaçantes sans doute par leur vivacité, mais permettaient aussi de varier et de revigorer les semis. L'humanité s'installe dans la monoculture; elle s'apprête à produire la civilisation en masse, comme la betterave. Son ordinaire ne comportera plus que ce plat. (38-39)

Lévi-Strauss utilise ici le vocable "monoculture" dans son double sens civilisationnel et agraire pour non seulement montrer le rapport entre l'assimilation culturelle et la monoculture agraire, mais surtout les dangers de la promotion d'une culture unique, quelle qu'elle soit. La monoculture inaugure la fin progressive, mais sûre de toute vie; dans le sens où elle favorise le déni complet du colonisé. La culture de celui-ci, son savoir-faire et les moyens de son rapport à son environnement sont totalement reniés. *Monocultures of the Mind*, le titre de l'essai de Vandana Shiva, est révélateur à ce sujet. En rapprochant monoculture et esprit, l'éco-féministe indienne établit implicitement le lien et les potentielles mutuelles influences entre la monoculture et la pensée unique.

Dans *La Terre du café,* le petit-fils évoque l'assimilation culturelle dont les Camerounais ont fait l'objet. L'une des marques de cette aliénation culturelle est l'uniformisation linguistique. En effet, à l'arrivée des Occidentaux, la région de l'ouest du Cameroun dont il est particulièrement question dans la nouvelle comptait plusieurs langues. Mais les Français, à travers leur politique d'assimilation, ont essayé de bannir toutes les langues autochtones au profit du français. L'un des effets directs de ce génocide linguistique[2] encore perceptible dans la société camerounaise contemporaine est que les jeunes, éduqués à l'école française, le plus souvent ne savent plus parler leur langue maternelle et toute communication avec ceux que l'on nomme "illettrés" est rompue. Le narrateur plaint la politique française de l'uniformisation linguistique:

> Non, je le sais: c'était la culture française qui éteignait la multitude de nos cultures sous le manteau ennuyeux de son assimilation jacobine [...]. C'était surtout la langue française qui éteignait le riche grenier de nos langues, la dizaine de langues Bamilékés par exemple, pour lui imposer sa rébarbative singularité. (16)

Le système colonial français vise l'uniformisation, et par conséquent détruit toute idée de diversité. Ce procédé d'homogénéisation, malheureusement ne se limitera pas à la seule sphère culturelle, elle s'étendra aussi au domaine environnemental et agricole en particulier.

La critique de la monoculture est au centre de la nouvelle de Nganang. Le narrateur décrit la violence de la colonisation de la terre "comme une vague violente qui couvre la terre vive de l'eau de son tsunami" (15). Il poursuit,

> Je voyais soudain notre village se vider de ses couleurs chatoyantes, de son million de bourgeons miraculeux, de son infini de variétés de vie, pour ne laisser place qu'à l'infini monotone de collines couvertes de champs de café. (15)

---

[2] La nouvelle aborde la perte de la diversité écologique et linguistique. De même que de nos jours on parle d'espèces de plantes ou d'animaux en voie de disparition, il existe également des langues fortement menacées d'extinction.

Par ces mots, le petit-fils déplore la disparition de la biodiversité au profit d'une uniformité périlleuse. La monoculture apparaît aussi comme une pratique négationniste. Comme l'explique le narrateur de *La Terre du café*, à l'arrivée des colonisateurs, les autochtones avaient leurs pratiques agricoles qui prônaient la diversité. Les colonisateurs n'ont pas établi des plantations sur des terres vierges. Ils ont organisé et encouragé la destruction des plantations existantes. Cet acte constitue en lui-même une négation d'un système culturel qui établit des modalités du rapport de l'homme avec son environnement. L'agressivité et la monotonie de la monoculture décriées par Nganang sont stigmatisées par Shiva: "Monocultures are in fact a source of scarcity and poverty, both because they destroy diversity and alternatives and also because they destroy decentralised control on production and consumption systems" (6). Les colonisateurs ont persuadé les natifs de transformer leurs champs en plantations de café. Les conséquences de cette conversion sont nombreuses et les principales sont la destruction de la diversité écologique et la dépendance des natifs par rapport aux colonisateurs.

Les plantations de café ont substitué les champs de divers produits locaux servant dans l'alimentation des populations. Or le café est une plante qui est complètement étrangère et aliénante pour les locaux: ils ne consomment pas le café. Paradoxalement, le café occupe désormais le point central de leur vie:

> C'est qu'il sera entré dans nos vies, le café, avec la brutalité du colon, aura réorganisé nos foyers, restructuré nos villages, redéfini nos priorités, et mis dans nos rêves, la seule révélation de sa plante verte, à qui seule nous avons donné la liberté de pousser sur nos terres. (18)

Le narrateur explique la puissance de la colonisation écologique. Une seule plante, tel un envahisseur, a réussi à démanteler tout le système d'un groupe social. Ce qui est le plus marquant c'est que le colonisé ou ancien colonisé ne joue qu'un rôle infime dans la filière du café. Dans son article "Cultivating Famine: In Search of the Miracle Seed," Azucena Cruz analyse les dérapages d'une agriculture qui cherche à se défaire de son rôle premier de production de nourriture pour assouvir les besoins vitaux de l'homme. Cruz estime qu'une telle agriculture court inéluctablement le risque d'anéantir les populations locales et leur environnement:

> [I]t is when we begin to see agriculture as something that is more than just a natural resource that fulfills our most fundamental needs — more specifically, when it becomes an issue of productivity and income or when the foods we eat become politicized — that it becomes a tool of power. The power that it creates is the power to control the landscape, to decide what kind of crops are harvested, in what way and in what quantity; the power to control who receives the benefits (in the broadest sense of the term) of the food produced and the income accumulated from crop sales; and, lastly, the power to shape and reshape cultures and nations in a multitude of ways. (43-44)

Cruz relève les caractéristiques aliénantes de la monoculture intensive. Celle-ci transforme l'environnement et rend les populations locales étrangères sur leurs propres terres qu'elles ont du mal à reconnaître. De plus, ces locaux, pourtant à la base de la filière caféière, sont complètement exclus du système. Le produit de leur labeur est entièrement destiné à l'exportation, ce qui signifie que sur le plan local, il ne contribue pas à l'alimentation. Par ailleurs, les prix de vente du café sont fixés encore par les Autres. Le narrateur constate: "tout le pays Bamiléké[3] était devenu un pays du café. [...] c'est toute l'Afrique qui avec la colonisation avait limité l'étendue de ses rêves à la dimension d'une tasse au petit-déjeuner des Français" (Nganang 30).

Certains auteurs africains ont utilisé la métaphore de la colonisation par les plantes pour montrer les méfaits du colonialisme sur l'environnement des colonisés. C'est le cas de Zakes Mda[4] et Ngugi Wa Thiong'o

---

[3] Bamiléké est le nom de l'ethnie principale de la région de l'ouest du Cameroun.

[4] Le Sud-Africain Zakes Mda utilise l'image de la plante colonisatrice dans son roman *The Whale Caller*. La tulipe, introduite par les Hollandais en Afrique du Sud supplante les colonisateurs eux-mêmes. Elle règne en maîtresse absolue dans les jardins, au détriment des plantes natives. Le narrateur décrit: "Tulips flower in spring, but these have developed erratic habits. They blossom any season they fell like blossoming, and they do it all at the same time, upstaging every other plant in the wild garden. And when they have decided to bloom,

dont les textes respectifs *The Heart of Redness* et *Petals of Blood* montrent comment les colonisateurs ont introduit de nouvelles plantes en terres colonisées pour détruire les variétés autochtones déjà présentes. Dans son analyse de *The Heart* of *Redness* et de *Petals of Blood*, Laura Wright conclut que

> the destruction of indigenous plants and their displacement by alien species can be read on a very literal level as indicative of the decrease of biodiversity that results from the introduction of nonnative species, but such an instance can be read as a cultural metaphor. (22)

Le viol de la terre des Bamilékés l'a transformée en une espèce de *pharmakon* différé, c'est-à-dire que cette terre n'a jamais été à la fois remède et poison. La caractéristique thérapeutique de la terre s'est mutée en toxine après sa transformation. Racontant les actes de son grand père, le petit-fils montre le côté palliatif de la terre: "Je Te vois mettre la terre dans Ta bouche et m'en donner à manger, tandis que Ton regard me dis que la terre n'empoisonne pas, mais guérit" (Nganang 10). La terre ici est sacralisée car présentée comme une panacée aux maux de l'homme. Mais plus loin dans le texte, on comprend que par le café, la terre tuait ses fils à petit feu, d'où le constat douloureux du petit-fils: "Je comprenais que c'était la richesse de notre terre qui faisait notre damnation" (14). Le décryptage de l'emploi du vocabulaire médicinal dans le texte est important pour comprendre la valeur de la terre sur la vie (celle de la terre elle-même et des hommes).

Les dégâts de l'éco-colonialisme ne se sont pas limités simplement à l'environnement. Ils se sont étendus sur le vivant en général en touchant la vie des hommes. C'est cette colonisation des corps humains qui est

---

sometimes after hibernating for three years without a peep of colour from them, they are relentless. They spread all over the garden and are not deterred by the wild shrubs and grasses and the prickly pear and other cacti that otherwise reign supreme in the garden" (105). Ce passage symbolise le pouvoir suprême des tulipes sur les autres plantes. L'on peut aussi noter l'idée que le colonialisme est un système qui a une propension au déguisement car il continue d'être à l'œuvre quand bien même il donne l'impression d'être mort comme les tulipes en hibernation. Cette tactique de camouflage cache mal l'hypocrisie d'un système caractérisé par l'ambivalence et la duplicité.

nommée le "bio-colonialisme."[5] L'éco-colonialisme et le bio-colonialisme sont les branches d'un même arbre. Ce sont des tendances du colonialisme qui, chacune à son niveau, tente d'asseoir l'idéologie dominante. L'éco-colonialisme vise particulièrement l'environnement humain, et dans le cadre de la nouvelle *La Terre du café*, il se concentre essentiellement sur l'agriculture. Or le principal but de l'agriculture est de sous-tendre la vie des hommes. Le bio-colonialisme peut prendre plusieurs formes et passer par différents biais. Il peut par exemple se servir de l'éco-colonialisme pour atteindre ses objectifs. C'est pour cette raison que le bio-colonialisme tel qu'abordé dans ce travail a une connotation indirecte. Il n'est pas question d'actions directes sur les corps colonisés mais les préjudices subis par les corps colonisés viennent en partie de la monoculture et de ses conséquences.

La politique française dans les colonies a influencé toutes les formes de vie. On a l'impression que les colonisateurs ont, à un moment donné, conféré leur pouvoir à une seule plante. Celle-ci s'est chargée de pénétrer jusque dans la peau des colonisés. A ce niveau, il est possible d'entrevoir une forme de biopouvoir tel qu'analysé par Michel Foucault. Dans *Histoire de la sexualité*, Foucault présente l'un des axes du biopouvoir en ces termes :

> centré sur le corps comme machine: son dressage, la majoration de ses aptitudes, l'extorsion de ses forces, la croissance parallèle de son utilité et de sa docilité, son intégration à des systèmes de contrôle efficaces et économiques, tout cela a été assuré par des

[5] Dans son ouvrage *The Global Genome* publié en 2005, Eugene Thacker définit le bio-colonialisme comme suit: "a term that has been used frequently in describing the practice of sampling genetic material for potential genes of value" (134). En effet, la controverse autour du projet financé par le gouvernement américain sur la constitution d'une base de données génétique de certains groupes ethniques a favorisé l'expansion de la notion de bio-colonialisme. Les peuples indigènes à travers le monde, les Amérindiens et les Aborigènes en particulier ont considéré un tel projet comme le bio-colonialisme. Ces communautés, déjà affaiblies par la colonisation qui les a dépouillés de leurs terres et leurs cultures se sont résolues à combattre ce système qui souhaite s'infiltrer au cœur de leur vie en s'accaparant leurs gènes. Disposer des données génétiques d'un peuple dans une base de données signifie posséder la vie de ce peuple, d'où la possibilité de la contrôler.

> procédures de pouvoir qui caractérisent les disciplines: anatomo-politique du corps humain. (183)

La vie (*bios*) est au cœur du biopouvoir. La colonisation, par sa politique, a le pouvoir sur la vie: pouvoir de donner ou d'ôter la vie. L'assimilation française a ôté la vie (environnementale et humaine) dans la colonie pour soutenir la vie en métropole. La colonisation française chez les Bamilékés a fabriqué une nouvelle génération d'hommes: les hommes de café. Le café est désormais cette autorité impersonnelle qui régit la vie et les corps des natifs.

Le grand-père du narrateur comme ses contemporains est devenu "esclave" d'une plante qu'il ne mange pas. Une fois affamées par la monoculture, les populations sont devenues vulnérables. Comme le dit l'adage, "un homme qui a faim n'est pas libre." Le corps du grand-père est la preuve visible des effets du biopouvoir écologique:

> Tes mains traversées par Ta longue vie ont creusé des trous dans le sol; Ton regard transformé par la dureté de la nature; Ton corps déchiré par le café; Tes pieds aux orteils écarquillés à force de s'enfoncer dans le sol dur — pour témoigner de la violence de la vie sous la coloniale. (11)

Le café a marqué le corps du colonisé comme des maitres qui marquaient leurs esclaves au fer chaud. Le narrateur donne une image de l'intrusion de la plante dans toutes les parties du corps du colonisé:

> Et la plante, je sais, poussa ses bourgeons dans la tête de chacun. Elle transforma des mains en feuilles vertes livrées au vent. Elle transforma les corps en caféiers au tronc rassis. Elle s'incrusta dans des colonnes vertébrales et tordit des corps. Elle fit des pieds s'enfoncer au sol avec la dureté des racines caféières. (18)

Dans *La Terre du café*, le caféier devient la métaphore même de la colonisation qui, à la façon d'un rhizome, occupe du terrain, circule partout, enlace les éléments sur sa voie sans rien épargner. Le processus que décrit le narrateur ressemble à ce qui est appelé en terme médical la xénogreffe. La xénogreffe est une greffe entre deux espèces biologiquement différentes. Généralement, la xénogreffe se solde par le rejet de l'organe transplanté. Cependant, la greffe de la plante dans le colonisé est une union qui dure et qui conduit à la mort lente du colonisé par asphyxie. C'est un meurtre masqué en mort naturelle. Une fois ancrée dans le colonisé, la plante représente la colonisation des corps et des esprits des colonisés. Désormais, le colonisé porte en lui le "germe" de la colonisation. Le pouvoir colonial est transféré au colonisé qui va devenir son propre bourreau.

L'auteur s'étend considérablement sur l'assimilation culturelle dont les colonisés ont fait l'objet. Mais dans la nouvelle, le narrateur ne prend pas le déracinement linguistique comme une fatalité. La langue française qui est désormais sienne sert d'outil pour combattre toute forme d'oppression. Le silence des Français et des Africains sur les bavures coloniales est contrasté dans le texte par une "écriture écologique."

Marilyn M. Cooper promeut l'écologie de l'écriture dans un article "The Ecology of Writing" où elle propose une nouvelle vision et approche de l'écriture: "What I would like to propose is an ecological model of writing, whose fundamental tenet is that writing is an activity through which a person is continually engaged with a variety of socially constituted systems" (367). Loin d'être un acte solitaire, l'écriture doit entretenir avec la société des relations qui s'inscrivent dans la durée. Sa suggestion d'une écologie de l'écriture souligne le caractère dynamique entre les éléments. Il y a trois notions à retenir de l'analyse de Cooper: communauté, relations et action. Ces trois éléments se retrouvent tout au long de la nouvelle de Nganang où l'histoire individuelle se transforme en histoire collective et plus encore, où la nature rencontre la culture.

Lorsque l'on voit la lettre du petit-fils à son grand-père, on s'attend à ce qu'elle ait un ton très intimiste et surtout qu'elle aborde des problèmes spécifiques entre le destinateur et le destinataire. Certes, la lettre revient sur l'histoire de l'enfance du petit-fils avec son grand-père, mais ce passé familial se confond avec l'histoire de tout un peuple. En faisant resurgir ses souvenirs, le petit-fils fait renaître l'histoire du peuple Bamiléké en particulier et de la colonisation en général. C'est à ce niveau précis que le schéma écologique de l'écriture de Patrice Nganang se dévoile par le croisement entre l'histoire individuelle et l'histoire collective. La lettre devient un point d'intersection entre l'histoire et l'Histoire. L'histoire du petit-fils et de son grand-père n'est pas celle d'individus isolés, mais celle de tout un peuple. La dimension communautaire et relationnelle, étant l'un des points importants de l'écologie, est soulignée dans cette nouvelle épistolaire. L'écologie en tant que science met l'accent sur les relations entre différents éléments. Elle prône l'entretien de l'interdépendance entre ces éléments et par conséquent le respect de chacun de ces éléments pour une meilleure vie de tout le système.

Par ailleurs, le narrateur dissout toute barrière entre la nature et la culture en rapprochant la terre et l'écriture. L'élément naturel représenté par la terre et le culturel qui est l'écriture possèdent des modes d'expressions similaires: l'écriture ou la terre "crie en silence" (24). La fusion des deux éléments se concrétise dans cette formule du narrateur: "La terre est meilleur écrivain" (14). Cependant, il faut noter que la terre et l'écriture peuvent être soit des antidotes, soit des substances toxiques, cela dépend de l'usage que l'homme en fait. Le petit-fils entend redonner à la terre et à l'écriture leurs attributs vivificateurs. La terre donne la vie aux plantes, aux animaux, aux hommes, tandis que l'écriture fait vivre des idées. Dans le cadre de la colonisation décrite par le petit-fils, la terre et l'écriture sont devenues meurtrières. L'écriture coloniale sélective et donc mensongère, tue une fois de plus tout un peuple et son histoire par l'omission choisie.[6] D'autre part, la terre de la colonie, jadis mère nourricière a été transformée en mère qui commet l'infanticide.

La nouvelle traite aussi d'une association intéressante: celle de la terre et du sang des hommes. Le petit-fils se remémore les conséquences meurtrières engendrées par les plantations de café. Ainsi, les victimes qui voulaient sauver les plantations de café contre les actes pyromanes, y ont laissé leur vie en "rougi[ssant] de leur sang la terre du café" (22). Désormais, la couleur rouge de la terre des Bamilékés rappelle la mort de leurs fils. De cette fusion entre la terre et le sang, il est ressorti plus d'une explication. D'une part, la terre, mélangée au sang des fils de la région est un pacte qui scelle l'attachement entre les Bamilékés et leur terre. D'autre part, le sang qui entre dans la terre du café est une marque

[6] La nouvelle *La Terre du café* fait référence à la colonisation écologique et au génocide Bamiléké perpétrés par la France. Cependant, les livres d'histoire français et même camerounais choisissent de taire cette partie de l'histoire. C'est ce trou de mémoire volontaire que Nganang entend combler dans son texte.

que le caféier est une plante vampire qui se nourrit du sang humain. En dépit de ces différentes interprétations, il n'en demeure pas moins que l'alliance entre la terre et le sang est un signe de la dissolution des lignes de démarcation entre l'homme et la nature.

Dans sa lettre, le petit-fils mentionne que la communication a été interrompue entre lui et son grand-père, du fait que l'un et l'autre utilise des moyens de communication différents. Ajouté à ces différences linguistiques et de mode de communication, la lettre est adressée à un défunt. À quoi servira cette lettre qui semble un objet de dialogue de sourds? Le narrateur donne la réponse à question à la fin de sa correspondance:

> Voilà pourquoi cette lettre que je T'écris, je viendrai la planter sur Ta tombe, la prochaine fois où je retournerai au Cameroun. Pas afin qu'elle jaillisse de terre, pousse, emplisse l'univers de ma rage, car à la différence des plantes, le papier se décompose sous la pluie; pas afin qu'elle parle, crie, hurle, habille l'univers de mon courroux, car à la différence des hommes, l'écriture crie en silence: afin que dans le silence des racines de la terre du café, elle dissémine mes mots avec l'eau qui pénètre jusqu'au cœur de la vie, donc dans la largeur éparpille de la chair du monde, elle inscrive la volonté de la France de se rebiffer devant ses crimes coloniaux; afin qu'elle marque dans les veines de notre sol notre conscience de l'appauvrissement de notre vie que fut et sera toujours la colonisation. Afin donc que lors des prochaines pluies, notre terre réveille mes mots Tiens avec son parfum pour dire la violente vérité de ce que Tu as vécu, et colle en boue aux pieds de tous les passants la mémoire de notre douleur Tienne. (24)

Le narrateur souligne aussi le mélange de l'eau et de l'écriture. Ces deux éléments sont à la base de la vie. Mis ensembles, ils constituent la sève, c'est-à-dire la substance nutritive qui insufflera la vie au monde. La terre et l'écriture ont aussi en commun une vie éternelle. L'écriture est cette plante d'une nouvelle nature que le petit-fils utilise pour redonner vie à l'histoire véritable. Cette écriture organique qui privilégie le dynamisme à travers l'intégration des éléments entre eux, la mobilité de ces éléments et enfin leur vie, subvertit l'écriture mensongère de la France. Le narrateur ici récupère l'écriture et la langue héritées de la colonisation pour résister aux perversités du régime colonial, en l'occurrence une plante et une écriture assassines. Finalement, la terre n'est pas que cette matière apparemment sans vie que l'on piétine, elle constitue la mémoire d'un peuple qu'elle tâche de transmettre à tous ceux qui entrent en contact avec elle. Par ce rôle, la terre ressemble étrangement à la langue, support incontournable de l'écriture que Ngugi Wa Thiong'o a défini comme "the carrier of culture."

En conclusion, la nouvelle *La Terre du café*, présente une imbrication entre le culturel et le naturel, tant dans les idées qu'elle révèle que sur le plan du style. Cette pollinisation croisée entre la nature et la culture donne l'impression que le texte de Nganang est un site où "the networks of language and culture" rencontrent "[the] networks of the land."[7] Loin d'être une pratique du passé, l'éco-colonialisme est encore manifeste dans de nombreux pays. Dans le cadre de l'Afrique, les multiples coopérations entre la Chine et plusieurs pays africains perpétuent la tradition éco-coloniale qui voudrait que l'Afrique soit une grande plantation pour les puissances économiques. Il est donc important pour les pays africains d'être vigilants par rapport aux accords et à la nature de leurs relations avec l'Autre. Comme le dévoile la nouvelle, l'abus que l'homme cause à la terre se répercute dans tous les domaines de la vie de celui-ci, d'où l'importance pour l'homme de comprendre et de toujours considérer son interdépendance avec les éléments qui l'entourent.

---

[7] Campbell, cité par Lawrence Buell dans *The Future of Environmental Criticism* (10). Buell explique ici l'hybridisation entre la nature et la culture.

Ouvrages cités

Armbruster, Karla, and Kathleen R. Wallace. *Beyond Nature Writing: Expanding the Boundaries of Ecocriticism*. Charlottesville: University Press of Virginia, 2001.

Buell, Lawrence. *The Future of Environmental Criticism: Environmental Crisis and Literary Imagination*. Malden: Blackwell Pub–lishing, 2005.

Cooper, Marilyn. "The Ecology of Writing." *College English* 48.4 (1986): 364-75.

Crosby, Alfred. *Ecological Imperialism: The Biological Expansion from Europe, 900-1900*. Cambridge: Cambridge University Press, 1989.

Cruz, Azucena: "Cultivating Famine: In Search of the Miracle Seed." *Ecoscapes: Geographical Patternings of Relations*. Ed. Gary Backhaus and John Murungi. Oxford: Lexington Books, 2006. 38-55

Foucault, Michel. *Histoire de la sexualite. T.1. La volonté de savoir.* Paris: Gallimard, 1976.

Knobloch, Frieda. *The Culture of Wilderness: Agriculture as Colonization in the American West*. Charlotte: University of North Carolina Press, 1996.

Lévi-Strauss, Claude. *Tristes tropiques*. Paris: Plon, 1955.

Mda, Zakes. *The Heart of Redness*. New York: Farrar, Strauss and Giroux, 2000.

______. *The Whale Caller*. New York: Farrar, Strauss and Giroux, 2005.

Nganang, Patrice. *La Terre du café. Lettre à mon grand-père à propos de la France qui ment*. In *Dernières nouvelles du colonialisme* (Collectif). La Roque d'Anthéron: Vents d'Ailleurs, 2006. 5-24

Said, Edward. *Culture and Imperialism*. New York: Vintage Books, 1994.

Shiva, Vandana. *Monocultures of the Mind: Perspectives on Biodiversity and Biotechnology*. New York & Penang (Malaysia): Zed Books & The Third World Network, 1993.

Thacker, Eugene. *The Global Genome: Biotechnology, Politics, and Culture*. Cambridge, MA: MIT Press, 2005.

Wa Thiong'o, Ngugi. *Petals of Blood*. London: Heinemann, 1977.

Wagner, Eric. "A Land's Allegiance: Functions and Representations of Landscape in Post-Colonial Theatres." *Interdisciplinary Studies in Literature and Environment*, 16. 3 (Summer 2009). 569-89

Wright, Laura. *Wilderness into Civilized Shapes: Reading the post-colonial Environment*. Athens, GA: University of Georgia Press, 2010.

Stéphanie Posthumus

McGill University

Writing the Land/scape: Marie Darrieussecq's *Le Pays*

An acclaimed contemporary French writer, Marie Darrieussecq has illustrated the need for rethinking the relationship between the text and the world in light of contemporary theories of language, mind, and writing. Her novel, *Le Pays* (2005), brilliantly explores this relationship both on a thematic level — the land constructed and experienced as landscape by a decentred, cosmopolitan subject — and on a formal one — the use of the *autofiction* genre and the *dédoublement* of narrative voices. This article proposes to examine these elements from an ecocritical perspective while at the same time navigating the cultural differences that have largely marginalized ecocriticism within French literary studies.

In his seminal 1995 book, *The Environmental Imagination*, Lawrence Buell outlines four characteristics of the environmental text: 1) the nonhuman environment is presented as more than a framing device; 2) non-human interests are given as much attention as human ones; 3) humans are seen as accountable to the environment; 4) the environment is represented as a process rather than a given. While Buell attempts to lay a theoretical groundwork for ecocriticism that transcends cultural differences, it is not clear to what extent his definition of the environmental text can or should be applied universally. Take for example the term "environment" that is key to Buell's definition and at the heart of the more general environmental movement in North America. In France, the term *environnement* is not a perfect and exact translation of the English word "environment." In its original sense of that which surrounds the human, the term *environnement* retains a strong anthropocentric bent that is not as apparent in English. Recycled back into English during the 1960s, the use of the word *environnement* to mean the physical world has been viewed with suspicion and at times rejected outright (Serres). Moreover, the term "environmentalism" that has been derived from the term "environment" to mean a concern for the physical world does not have an exact equivalent in French. *Environnementalisme* remains a borrowed term from English while the expression *écologie politique* covers a different set of disciplines and discourses. What may seem to be a trivial linguistic difference reveals deeper cultural differences that can be traced to specific socio-historical factors.[1]

To further illustrate the need to contextualize the notion of an environmental text and more generally ecological reading, I will analyze French author Marie Darrieussecq's 2005 novel, *Le Pays*. A contemporary and relatively prolific French writer, Darrieussecq admits that her writing can be difficult at times as the story jumps from one idea to the next, one event to the next, one narrator to the next, leaving the reader to fill in the gaps ("Comment j'écris" 262). As Simon Kemp argues, these gaps result from Darrieussecq's attempt to represent the processes of thinking and thought in her writing ("Darrieussecq's Mind"). Despite this emphasis on the mind, Darrieussecq's literary experiments remain strangely rooted in the real world. While steeped in the ideas of the *Nouveau Roman*, these texts de not subscribe to the division between *l'aventure de l'écriture* and *l'écriture d'une aventure* (Ricardou). Instead, they explore the space between language and reality, pushing the reader to reflect on literature's role in society as well as on language's role in thinking about the physical world.

Darrieussecq's *Le Pays* tells the story of a young female writer, Marie Rivière, who, after living many years in Paris moves with her husband, Diego, and their son, Tiot, back to the region where she grew up. Back in her hometown, Marie Rivière begins work on a new novel to be entitled *Le Pays*, while also attempting to reconnect with her family and the places and memories of her childhood. The *mise-en-abyme* form – a novel about a novelist writing a novel – allows Darrieussecq to raise important questions about identity, nationality, land-

[1] These cultural differences have played a role in the delayed development of environmental history (Locher and Quenet) and environmental philosophy (Afeissa; Larrère) in France where similar questions about the use of the word *environnement* have been raised.

scape and belonging. It is this tension between literature as a closed network of texts, and literature as a way of bridging word and world that makes Darrieussecq's novel a particularly productive choice for an ecological literary reading.

While critics have astutely examined questions of identity, mind, narrative voice and the feminine body in Darrieussecq's work, the role of landscape has yet to be given much attention. Such an omission is surprising given the author's assertion that her writing begins with landscapes. Referring to her best-selling novel, *Truismes* (1996), Darrieussecq explains: "Quand je commence à rêver sur un livre, en général, je vois d'abord des paysages. Je savais qu'il y aurait la Seine, un Paris apocalyptique, ruiné" ("Comment j'écris" 257). This same idea can be found in *Le Pays* when Marie Rivière asserts that she wants to write a novel that starts "comme ça, paysages et questions" (85). Using landscapes to begin a novel does not mean relegating the text to a set of images. As the novel clearly illustrates, landscapes are intimately related to real, geographical places. Marie Rivière decides to move back to where she grew up because the urban Parisian parks and squares do not offer her son, Tiot, a true experience of landscape. The sense of this word becomes clearer when she asserts: "Je voulais un paysage pour Tiot… Je voulais proposer un pays pour Tiot" (44). In other words, landscape is not some abstract, virtual image existing solely in the subject's mind. It is instead tightly bound up with specific places and in this case a country/land with which Marie Rivière maintains a complex and difficult relationship.

In French as in English, the words landscape/*paysage* and land/ *pays* are closely related etymologically, a fact that has been taken up by landscape theorists as a way of defining landscape. In his 1997 study, *Court Traité du paysage*, Alain Roger asserts that the land is the material that makes up landscape, the former being physical, natural, "brute matter" and the latter cultural, constructed, human. Literary critic, Michel Collot, bases his definition of landscape on a similar distinction between land and landscape: "Le paysage, c'est une certaine mise en forme du pays, qui permet de le saisir comme un ensemble" (193). In both these cases, landscape is defined as a distinct aesthetic, culturally informed experience of the land. Yet a closer reading of the landscapes in Darrieussecq's novel reveals the impossibility of establishing such a clear separation between the land as physical and the landscape as cultural.

A large part of the story takes place in Yuoangui, a barely disguised fictional counterpart to *le Pays basque* where Darrieussecq herself grew up. Yuoangui is bordered on the west by the ocean, on the east by a mountain range on the north and the south by other European countries.[2] In some sense, the landscapes described in the novel can be mapped onto real geographical places, thus closing the gap between the text and the world. And yet Darrieussecq carefully and constantly reminds the reader that she is not writing about a real country, no matter how real the landscapes may seem. While the geography of Yuoangui is identical to that of *le Pays basque*, it is a fictional country that recently acquired political independence, with its own official language and national literature. One might too quickly assume that Darrieussecq is pushing for a similar status for her own birthplace, *le Pays basque*. Yet as the story unfolds, it becomes clear that Darrieussecq is more interested in exploring the (often ambivalent) connections that a writer might feel towards the land and landscapes of her birthplace.

The most common landscapes in the novel, those of the mountaintop and the seaside, seem familiar to the reader in that they fall into the traditional category of landscape as a wide expanse of land seen from a distance. Bordered by the ocean and a mountain range, Yuoangui has the particularity of being a very small country with great natural beauty. But the novel does not let the reader forget that such "natural" beauty is both "un cliché et une nostalgie" (65) often used to promote tourism. Marie Rivière highlights the fact that these experiences of the land's beauty are carefully crafted in order to satisfy the city dweller's desire for "beauté, bon air, mer" (43). For example, during one of her many road trips along the coast, she remarks:

> La falaise brisée, les rochers jumeaux, la ligne d'horizon sur mer et montagne, le fleuve et l'ancienne frontière: beaucoup de cartes postales sont prises de la Corniche. Les cartes routières sont marquées 'point de vue' à son emplacement. (80)

---

[2] Darrieussecq explains that the word "Yuoangui" was a "mot générique" that she invented when writing *Naissance des fantômes* to mean a place where all the immigrants and outcasts of the world live ("Des livres" 112). Given the long conflicts between the Basque people and the French government, it is not surprising that she might choose to use the same word in *Le Pays*.

Such commercialisation of the natural landscape is not however criticized. In fact, Marie Rivière chooses to stop over and over again at the spots marked "scenic" in order to situate herself with respect to the land: "Finalement la vie aurait pu passer ainsi, d'aire de pique-nique en aire de pique-nique, de point de vue en point de vue" (184). She returns to the seaside as a way of remembering the landscapes of her childhood but also to come to terms with the fact that much has changed.[3] Resituating herself with respect to the seascape is one way of reconnecting with the land.

This connection involves a form of undoing the self in the case of the seascapes. As a recurring element in Darrieussecq's work, the sea represents a space where the subject often intensely experiences (being in) the world. This is true of Marie Rivière in *Le Pays* on a gray autumn day:

> Ce que vous êtes à l'intérieur se retrouve à l'extérieur. Vos molécules se mélangent au ciel et à l'eau, la solitude se diffuse. Les mots et les choses s'écartent, la pensée ne suit plus, les signes se désamorcent ; et le moi devient une grande béance pleine d'eau salée. (84)

Such a dispersion of self evokes an earlier fetal environment, suggesting that Marie Rivière is searching for a way of being (in) the landscape that does not necessarily have recourse to language or thinking.

At the same time, the novel acknowledges the role that landscape plays in the construction of national identity. While Marie Rivière experiences the seaside landscape as an undoing of the subject/object division, she makes note of the fact that "le pays est fédéré par le paysage vu depuis la Corniche" (80). As French historian François Walter

---

[3] Anthropologists Yvan Droz and Valérie Miéville-Ott distinguish between "le paysage pratiqué" and "le paysage remémoré" while recognizing that the memories of a landscape inflect later experiences of it. It is this overlap that brings out the disparity that Marie must learn to navigate: "Elle était rentrée au pays mais elle avait beau l'arpenter, user semelles et pneus contre sa croûte et dans ses sables, tout lui échappait. Le pays n'était pas le pays du souvenir. Tout était semblable et tout était différent, les temps ne se juxtaposaient pas, les paysages laissaient du jeu" (76).

explains, nations and national sentiment arise from the identification with particular landscapes.[4] Struggling to identify with the new country as a political entity, Marie nevertheless recognizes that the landscapes are the bridge between her and some sense of belonging to the land. Her identity then is less defined by the political than by personal encounters of specific places.

While the seascapes allow Marie to experience place intimately, the mountaintop views remind her that she is a part of a larger global world. Climbing mount Glyph, one of her first activities upon returning home, she hopes to discover "comment on fait, avec ce pays" (48). Yet once she reaches the summit, her mind wanders. When the green leaves of the trees below remind her of the already changing chestnut trees in Paris, she admonishes herself: "*Reste ici et maintenant*" (49). She then adds: "Le *moi* est un vaisseau spatial, capable de relier des univers, de rabattre les unes sur les autres des galaxies lointaines" (49). The body experiences the local landscape while the self balloons out to countries of similar geographic shape: "Océan à l'ouest, forêt au Nord, montagnes à l'est et Sud jaune. Une très petite Chine. Une grande île Lofoten" (49). Comparing China, a Norwegian archipelago and a small country the size of *le Pays basque*, Marie Rivière translates the experience of a local place into a larger global space, outlining her search for a cosmopolitan identity that does not reduce her work to the regional or even the national.[5]

This bird's eye view afforded by the mountaintop contrasts with the dissolution of the self that is brought about by experiencing the country's seascape. Marie Rivière admits her desire to encompass the country "d'un seul coup d'œil, un coup d'œil circulaire, comme une île… de le contenir, le faire sien" (73). Yet at the same time she recognizes the such distance can only be temporary as the need to contain the land quickly gives way to a sense of being "débordée" (73). Land/

---

[4] See also Pierre Donadieu's theory of landscape: "Les identités sociales peuvent s'enraciner dans les paysages en différenciant les groupes sociaux au même titre que les autres signes symboliques" (122).

[5] The relationship between such a small country and the global planet is an important subtext in *Le Pays*. As Darrieussecq explains, "Le paradoxe *du Pays* c'est que c'est un livre universaliste quant aux petits pays. Ce n'est pas du tout la défense d'une patrie. C'est un livre très utopiste qui imagine une planète faite de petits pays" ("Marie Darrieussecq. Écrivain").

scapes are thus the locus of a reciprocal process of opening up and out, the observer to the world and the world to the observer.

This is particularly evident in the novel's opening scene. Jogging early in the morning, Marie Rivière describes fragmented, moving images of the road and its surroundings that hardly seem to constitute a landscape: "Étape par étape, je ralliais des jalons, un arbre, un panneau, un champ. [...] Le macadam reculait sous ce qui avançait à ma place. Les bas-côtés, les fossés, les arbres et les collines se déplaçaient" (12). This constantly changing point of view calls into question the traditional idea of landscape as a coherent set of objects seen and constructed by an immobile observer. As the narrative continues, the lines between subject and object, between land and landscape are erased:

> J/e devenais la route, les arbres, le pays. S'absorber dans, absorber le paysage, c'était une partie de la pensée, une partie de l'écriture. Se remémorer le monde, une heure de rang, en courant. Le pays m'entourait, ce paysage familier qui devenait tous les paysages. (14)

Through this experience of landscape, Marie finds a creative space where she can begin writing *Le Pays*.[6]

Coming back to Buell's definition of the environmental text, one might ask if Darrieussecq's novel fits the given criteria once the shift is made from environment to landscape.[7] First, it is clear that landscapes

---

[6] In rather poetic fashion, French philosopher Michel Serres describes the writer as similar to the farmer, composing the landscape page by page and *pagus par pagus* respectively. He attempts to bring the world and the text together by this comparison: "Les pages ne dorment pas dans le langage, elles tirent leur vie des pagi: du paysage, de la chair et du monde" (262).

[7] This shift is key to recontextualizing the environmental text. While Buell draws on the force of grass-roots environmental movements that sprouted up in great number in the United States over the 1960s and 1970s in his assertion of the need for an ecological literary theory, such a move does not fit with the French cultural context. One of the premises of the present article is that the mobilization around concepts and theories of landscape in France over the last twenty years is a more appropriate framework for articulating a French *éco-critique*.

represent more than a framing device in the novel and that they are described and experienced as a process rather than a given. Second, they illustrate the inherent connections between the self and the physical world but also more generally point to the politics of national identity. In these respects, the novel does fulfill the function of the environmental text as outlined by Buell; yet it would be a mistake to limit an ecological reading of Darrieussecq's novel to these criteria as much of the novel's potential for pushing the reader to rethink the relationship between the text and the world lies in its innovative formal elements.

While Buell offers little in the way of analysis of formal elements, earlier ecocritical work makes some interesting attempts at classifying genre. In his pioneering 1974 study, *The Comedy of Survival*, Joseph Meeker argues that comedy is a more "ecological genre" than tragedy because its creates room for ambiguity, laughter, and hope in the face of ecological catastrophe. Despite the seemingly reductionist conclusion, Meeker explores the important question of what an ecological narrative genre should/could look like. While Darrieussecq's novel does not fall easily into either categories of comedy or tragedy, it does explore how a new genre like *autofiction* might fit with a rethinking of the social function of literature, a question that is at the heart of most ecocritical literary theories.

As already noted, *Le Pays* tells the story of a writer writing a novel about landscapes and country. This *mise-en-abyme* form highlights the role of writing as production rather than as a means of expressing some prior given reality. While Marie Rivière is deeply attached to the country's landscapes, her sentiment towards the country as a political entity remains ambivalent. She does not speak the language nor does she identify with the country's national literature, characterizing herself instead as a European writer (124). However, in order to write a novel about *le pays*, Marie Rivière needs to return to *le pays*, revealing the interdependent relationship between experiencing a place and writing about it. This relationship is renforced by the use of *l'autofiction* in Marie Darrieussecq's work.[8] Although the main character in *Le Pays* is

[8] Entitled *Moments critiques dans l'autobiographie contemporaine. Ironie tragique et autofiction chez Georges Perec, Michel Leiris, Serge Doubrovsky et Hervé Guibert*, Darrieussecq's doctoral thesis reveals her early interest in *autofiction* that she defines in a more recent article as "une 'fiction de soi' nourrie, comme en psychanalyse, par l'association d'idées et de mots" ("Je est unE autre").

called Marie Rivière and not Marie Darrieussecq, the parallels between the two are obvious.[9] When Marie Rivière describes her method of emptying herself and then trying to think and write from that position of emptiness (227-28), she echoes Marie Darrieussecq's explanation of her search for "l'absence à soi-même" as a starting point for writing ("Comment j'écris" 261). Moreover, both Marie Rivière and Marie Darrieussecq explain that absence and emptiness are not closed spaces filled by language and writing but rather they represent an openness into which elements of the real world flow and combine with images and phrases. As Darriessecq explains, her goal is still to "dire le monde" but to do this "il faut être dans une rêverie proche d'une sorte d'extase dans la langue, où on sort de son moi privé" (261). While the *mise-en-abyme* form has been analyzed in terms of the distance it creates between the text and the real world (a narrative world within a narrative world), the *autofiction* genre brings the two back into close contact as Darrieussecq uses fragments of her own life in the novel, teasing out the threads of the relationship between language/self/writing.

Another key characteristic of the novel's innovative form is the echo created by the two narrative voices. Using two font types allows Darrieussecq to move between the first person voice of Marie Rivière telling her own story and that of the third person narrator telling Marie Rivière's story. This *dédoublement* creates a narrative landscape that requires the reader to move between two perspectives, at times caught between the two, not always sure of who is speaking. While the first person narrative plunges the reader into Marie Rivière's (fragmented) consciousness, the third person narrator moves the reader into the position of observer. Yet the differences in the two voices are tempered as the third person narrator picks up threads from the first person narrative and begins weaving them into her own perspective. These two voices mirror the play of the *autofiction* that combines the author's (imagined) autobiographical elements with those of the fictional character, reminding the reader of the role that literature and language play in the (de)construction of subject identities.

---

9 Darrieussecq's last name means "a dry riverbed" while the main character's last name is "Rivière." Moreover, Darrieussecq does not hide the fact that *Le Pays* was written in the spirit of *autofiction*: "À part sans doute *le Pays*, je n'ai jamais écrit d'autofiction" ("Je est unE autre").

At the same time, the novel highlights the ways in which language can be reinvented in order to express multiple subject identities. The narrative *dédoublement* is mirrored on the lexical level by the use of the "j/e" in the novel rather than the usual "je" when the first person narrator observes herself from a distance. In the opening scene of the novel, Marie speaks of the "j/e" as she is jogging, letting her mind wander in order to find the space where writing can take place: "J/e ne pensais à rien. J/e courais. [...] Air, paysage, course. J/e ne pensais à rien et dans le rien perçaient les phrases, de plus en plus vite" (15). It is in this dynamic space of the autofictive "j/e" that the author's voice can be heard as she writes about local landscapes to which she is personally attached and yet detached as a well-travelled global citizen. As Catherine Rodgers explains, the play between fiction and reality characterizes all of Darrieussecq's work: "Elle transgresse constamment les limites entre la réalité et la fiction. Darrieussecq joue avec les similitudes entre sa réalité biographique et celle de sa protagoniste" (412-13). Playfully undoing the fixed subject positions of author, character and narrator, while at the same time revealing the instability of categories such as fiction and non-fiction, the *autofiction* genre represents an key element for thinking about a culturally and literary specific French *écocritique*.

Revisiting the ecological politics at the heart of Buell's definition of the environmental text, one might ask what other ecological readings of Darriessecq's novel are possible. Darrieussecq's novel includes as a subtext the effect of nuclear reactors on the landscape that Marie Rivière holds as central to her experience of (writing) the land. Taking deep breaths of fresh air in her garden one morning, Marie Rivière is reminded that even air, rain and dew can kill. She reflects on how the Chernobyl eco-disaster respected no national boundaries in Europe (64). She realizes that, despite its small size, isolated position, and incredible natural beauty, Yuoangui is part of a larger global ecosystem that is at risk: "Est-ce qu'ici, on est à l'abri? Est-ce qu'on est au cœur du monde, ou est-ce qu'on est à l'écart?" (194) Her comments elsewhere in the novel reveal her belief that all places and countries no matter how small are vulnerable to the risk of ozone holes, devastated forests and nuclear reactors. This sense of global ecological risk or "sense of planet" (Heise) does not however translate into an environmental politics. Marie Rivière's comments about the local nuclear reactor's effects on marine life and water temperature are not accompanied by a demand for change. Her husband happily surfs in the ocean next to the reactor's cooling system while their son plays on the beach. Aware of the risk, Marie Rivière seems to content to note the ecological issues without critiquing the nuclear energy industry.[10]

Such ambivalence may lead some readers to conclude that Darrieussecq's *Le Pays* is not "ecological enough." I would argue however that the quality of being ecological or environmental is not located in the text but rather in the literary critic's approach to the text. In this way, any novel can be read from an ecological perspective no matter what its themes are, from the anti-environmental remarks in Michel Houellebecq's *Les Particules élémentaires* and *La Possibilité d'une île* to the strong, pro-environmental message of Hervé Jaouen's *Que ma terre demeure* and *Aux armes zécolos*. This literary perspective allows for more difference and diversity in terms of the types of ecological politics and philosophies that have developed in various cultural contexts. As ecocritic Ursula Heise has often remarked, cultural, historical and political differences must be taken into account when developing an ecologically informed literary approach.

These differences mark the reception of ecocriticism in France where the *Nouveau Roman*, "l'ère du soupçon" and "l'auteur est mort" have only recently given way to a renewed interest in the self, history and the real (Blanckeman; Vercier). It is not surprising that French literary scholars Thomas Pughe and Michel Granger insist on narrative form in their analysis of American nature writing and opt for the expression "écrire la nature" rather than "écriture de la nature" (3). Such an emphasis on writing as a process acting on the natural world reminds the reader that language can never be a mirror for expressing the natural world. In the same way, the title of this paper, "writing the land/scape" rather than "landscape writing," is a way of emphasizing the active role that writing plays in creating relationships to land and landscape. This emphasis also reflects Darrieussecq's assertion about the need for new literary forms to help us imagine our rapidly changing world: "Je cherche à inventer de nouvelles formes, à écrire de nouvelles phrases,

[10] Historian Michael Bess offers a comprehensive explanation of why the French are hesitant to attack the large number of nuclear reactors in Europe (*The Light-Green Society*). Darrieussecq does not reveal her opinion one way or the other although the narrator in *Truismes* does include a tongue-in-cheek comment about her pig metamorphosis having something to do with the proximity of a nuclear reactor.

parce que c'est le seul moyen de rendre compte du monde moderne" ("Entretien réalisé par Becky Miller et Martha Holmes"). Writing thus becomes a way of better engaging with the myriad issues raised by today's globalized economy.

To conclude, it is worth emphasizing that Darrieussecq's work interests with some of the key issues at the heart of an environmental literary theory. First, her engagement with writing remains rooted in a desire to bring the reader to a new understanding of the material world:

> Le monde est aussi fait d'électrons, de microbes, d'ondes, de planètes... bientôt sans doute de clones, d'OGN, de nouveaux sons, de nouvelles odeurs... etc... Je participe au mouvement permanent des défricheurs. Je veux ouvrir des yeux sous les yeux des lecteurs, des oreilles sous leurs oreilles, une nouvelle peau sous leur peau. ("Entretien réalisé par Becky Miller et Martha Holmes")

By looking forward to future possibilities rather than trying to recreate some idealized past, Darrieussecq attempts to understand technology's effect on our conception of the human body, the world and the universe at large. Second, the emphasis on land and landscape in *Le Pays* acts as a counterbalance to the role played by wilderness in many ecocritical studies. As French historian François Walter explains:

> Aux États-Unis domine une conception de la nature sauvage et vierge, la *wilderness*. Au contraire, en Europe, la nature peut difficilement être séparée des aspects esthétiques, éthiques et patriotiques mobilisés pour différencier les nations. (264)[11]

Darrieussecq brings the culturally and nationally constructed elements of landscape clearly into focus. Third, writing represents a political act for Darrieussecq as it subverts dominant social discourses and creates new ways of thinking about the world outside the text:

---

[11] When areas were classified as "les réserves naturelles" in France, it was often in terms of their aesthetic beauty and their status as "patrimoine culturel" not in terms of their biodiversity or indigenous plant and wildlife species (Ford; Kalaora).

> En ce sens toute écriture exploratrice, novatrice, est politique: même apparemment éloignée du 'réel', des 'événements', elle fournit le langage moderne, elle bâtit les outils verbaux et mentaux qui permettent de penser le monde. ("Entretien réalisé par Becky Miller et Martha Holmes")

By exploring new narrative forms and insisting on the relationship between literature and the material world, Marie Darrieussecq's work fits squarely within the purview of a French *écocritique* that takes into account cultural specificities while also acknowledging the ways in which the literary text speaks to a larger public about global issues.

Works Cited

Afeissa, Hicham-Stéphane. *Éthique de l'environnement.* Paris: Vrin, 2007.

Bess, Michael. *The Light-Green Society: Ecology and Technological Modernity in France, 1960-2000*. Chicago: University of Chicago Press, 2003.

Blanc, Nathalie, Denis Chartier, and Thomas Pughe. "Littérature & écologie: Vers une écopoétique (Introduction)." *Écologie & politique* 36 (2008): 17-28.

Blanckeman, Bruno, Jean-Christophe Millois *et al. Le Roman français aujourd'hui. Transformations, perceptions, mythologies*. Paris: Prétexte, 2004.

Buell, Lawrence. *The Environmental Imagination Thoreau, Nature Writing and the Formation of American Culture.* Cambridge, MA: Harvard University Press, 1995.

Collot, Michel, ed. *Les Enjeux du paysage*. Bruxelles: Ousia, 1997.

Darrieussecq, Marie. "Comment j'écris." Interview with Jean-Marc Terrasse. *La Création en acte.* Eds. Paul Gifford and Marion Schmid. Amsterdam: Rodopi, 2007. 253-68.

______. "'Des livres sur la liberté': Conversation avec Marie Darrieussecq." Interview with Jeanette Gaudet. *Dalhousie French Studies* 59 (2002): 108-18.

______. "Entretien réalisé par Becky Miller et Martha Holmes." December 2001. <http://darrieussecq.arizona.edu/fr/entretien2001.html> May 2011.

______. "Je est unE autre." *Écrire l'histoire d'une vie.* Sous la direction d'Annie Oliver. Rome: Edizioni Spartaco, 2007. <darrieussecq.arizona.edu/fr/collautofiction.doc> May 2011.

______. *Le Pays.* Paris: POL/Folio, 2005 (2007).

______. "Marie Darrieussecq. Écrivain." Interview with Thomas Pierre. May 2008. <http://www.euskonews.com/0549zbk/gaia54903fr.html> May 2011.

______. *Truismes.* Paris: POL/Folio, 1996 (2008).

Donadieu, Pierre, Michel Périgord, and Lionella Scazzosi. *Le Paysage. Entre natures et cultures*. Paris: Armand Colin, 2007.

Droz, Yvan and Valérie Miéville-Ott. "Le Paysage de l'anthropologue." *La Polyphonie du paysage.* Ed. Yvan Droz et Valérie Miéville-Ott. Lausanne: Presses polytechniques et universitaires romandes, 2005. 1-20.

Ford, Caroline. "Nature, Culture and Conservation in France and her Colonies 1840-1940." *Past & Present* 183.1 (2004): 173-98.

Heise, Ursula K. *Sense of Place and Sense of Planet: The Environmental Imagination of the Global.* New York: Oxford University Press, 2008.

Kalaora, Bernard. *Le Musée vert. Radiographie du loisir en forêt*. Paris: L'Harmattan, 1993.

Kemp, Simon. "Darrieussecq's Mind." *French Studies*. LXII.4 (2008): 429-41.

______. "*Homeland*: Voyageurs et patrie dans les romans de Marie Darrieussecq." *Nomadismes des romancières contemporaines de langue francaise.* Ed. Audrey Lasserre and Anne Simon. Paris: Presses Sorbonne Nouvelle, 2008. 159-66.

Larrère, Catherine. "Éthiques de l'environnement." *Multitudes*. 24.1 (2006): 75-84.

Locher, Fabien and Grégory Quenet. "L'Histoire environnementale. Origines, enjeux et perspectives d'un nouveau chantier." *Revue d'histoire moderne et contemporaine*. 56.4 (2009): 7-37.

Meeker, Joseph. *The Comedy of Survival. Literary Ecology and a Play Ethic*. New York: Scribner, 1974.

Pughe, Tom and Michel Granger. "Introduction." *Revue française d'études américaines* 106 (2005): 3-7.

Ricardou, Jean. *Nouveaux Problèmes du roman*. Paris: Editions du Seuil, 1978.

Rodgers, Catherine. "Aux limites du moi, des mots et du monde: Questions d'identité dans *Le Pays* de Marie Darrieussecq." *Le Roman français de l'extrême contemporain*. Ed. Barbara Havercroft, Pascal Michelucci and Pascal Riendeau. Québec: Nota bene, 2010. 403-22.

Roger, Alain. *Court Traité du paysage*. Paris: Gallimard, 1997.

Serres, Michel. *Les Cinq Sens*. Paris: Grasset, 1985.

______. *Le Contrat naturel*. Paris: Champs/Flammarion, 1990 (1992).

Vercier, Bruno, Dominique Viart, and Franck Evrard. *La littérature française au présent. Héritage, modernité, mutations*. Paris: Bordas, 2005.

Walter, François. *Les Figures paysagères de la nation*. Paris: Éditions de l'École des Hautes Études en Sciences Sociales, 2004.

Liliane Toss

North Greenville University

L'Environnement dans le dessin de presse de Plantu: étude sémiotique

Le dessin de presse reflète l'actualité de l'époque et en souligne les crises et les désarrois. La présente étude prend en considération des dessins du caricaturiste du *Monde*, Plantu. Il s'agit d'une analyse sémiotique qui retrace le parcours interprétatif suivi par le lecteur pour en déchiffrer la signification. L'analyse des moyens d'expression propres au genre textuel du dessin révèlent la prise de position du dessinateur à l'égard de problèmes tels que la crise de la faim, l'exploitation des ressources naturelles, les déchets industriels ou la pollution. Les informations transmises par la caricature reçoivent une nouvelle dimension rendant évident son message.

Les dessins de presse occupent une fonction narrative car ils transmettent — par le signe iconique — une valeur cognitive. C'est en partant de l'idée d'un faire-sens visuel que l'analyse sémiotique de ces dessins révèle des détails bien intéressants dépassant parfois de loin le sujet d'actualité auquel ils font écho.

Jean Plantureux, plus connu sous le nom de Plantu, est le caricaturiste du *Monde* et de L'Express dont le dessin est connu pour être "plus redoutable que bien des analyses" (Payot). Plus qu'un simple caricaturiste, Plantu est considéré comme un "éternel metteur en scène" (Payot).

Les dessins qui portent la griffe de Plantu sont pour la plupart destinés à transmettre un message. Nombreux d'entre eux traitent des problèmes de l'environnement et condamnent la pollution sous ses divers aspects. Or, comme tout artiste, le caricaturiste dispose d'un matériel spécifique susceptible d'attirer l'attention du lecteur. Pour parvenir à son objectif, le dessin de presse utilise des moyens d'expression qui lui sont propres comme entre autres, l'exagération de certains détails, les variations d'échelle "ainsi que les petits signes graphiques qui traduisent le mouvement" (Darras 10).

En vue de décrypter le message d'un dessin de presse, le lecteur doit être au courant de l'actualité. En général, les faits illustrés figurent dans la même page du journal ou de la revue; ils peuvent être indiqués dans l'éditorial, dans le titre principal d'un article ou au sommaire. Le lecteur ainsi mis au courant du contenu de l'actualité, rétablit un parcours interprétatif du dessin qui l'accompagne et en découvre la signification. "Comme tout écrit, une photographie, un dessin ou une bande dessinée présupposent un apprentissage du code de lecture, une 'connaissance' (de l'ordre du signifié) antérieure qui, seule, permet la 'reconnaissance'" (Courtés, *Du lisible* 248). Par conséquent, le dessin de presse occupe, par sa situation même, une place particulière, très souvent en relation plus ou moins directe avec le titre de l'article au sein duquel il figure. Le lecteur essaie alors d'en déchiffrer le code de lecture afin d'en révéler le signifié/message. En d'autres termes, nous, lecteurs, partirons de "ce que nos yeux voient (= le signifiant, l'expression)" pour parvenir à "ce que notre esprit comprend (= le signifié, le contenu)" (Courtés, *Du lisible* 252). L'analyse du dessin se poursuit à deux niveaux, celui de l'expression, des structures narratives de surface et celui, plus profond, du thématique.

Un regard attentif accordé à tout ce qui relève du signifiant, de la perception visuelle, entraîne une compréhension plus approfondie du signifié, déjà souligné, ne serait-ce que partiellement, dans le titre de l'article l'accompagnant. Or, le figuratif "ne saurait se replier en quelque sorte sur lui-même: il appelle nécessairement [...] une thématisation" (Courtés, *Analyse* 164). Selon Joseph Courtés, l'analyse sémoitique se donne comme objet la signification primaire qui pourra être appréhendée par un lecteur standard. Dans le cas du dessin de presse, la rentabilité d'une telle analyse enrichit l'interprétation en mettant en évidence le fonctionnement des divers éléments figuratifs. "Un élément du contenu d'un texte (verbal ou visuel) est dit figuratif lorsqu'il a un correspondant perceptible au plan de l'expression du monde naturel" (Everaert-Desmedt, "Le parcours" 130). A titre d'exemple, le concept de pollution est thématique mais ses conséquences sont d'ordre figuratif. L'illustration du concept par des motifs repérables tels que la fumée ou les points noirs de poussière amène le lecteur à partager les valeurs de l'énonciateur.

L'on se penchera alors, dans les pages à venir, sur quelques-uns des dessins de Plantu pour essayer de mettre en relation le plan de l'expression (signifiant) exploité par le dessinateur du *Monde*, avec le plan du contenu (signifié) auquel le lecteur est supposé parvenir. Le parcours génératif de la signification d'un dessin une fois restituée, la prise de position de Plantu quant aux problèmes de l'environnement, se précisera de façon distincte. Ainsi, aux informations sur l'actualité que le lecteur découvre en lisant son journal, vient s'ajouter le point de vue du dessinateur.

L'itinéraire méthodologique que nous adoptons pour l'analyse du dessin allant du signifiant au signifié, il nous faudra poser la question de Courtés: "Quels sont les éléments visuels — relevant de la "matérialité" de la perception — qui, dans une bande dessinée, par exemple, seront sélectionnés parce que pertinents à la signification?" (*Du lisible* 247). Un premier dessin (Figure 1) illustre le thème de la pollution en France. Les journaux soulignent la menace en indiquant qu'en une seule journée, "le niveau 2 est dépassé dans treize départements" (*La France* 151). Il paraît que les citoyens de l'Hexagone sont condamnés à vivre dans une ambiance polluée où tout, y compris l'air qu'ils respirent, est contaminé. Examinons d'abord le plan de l'expression dans le dessin. Au niveau figuratif, il s'agit de deux formes graphiques de deux personnes, occupant chacune la moitié du dessin. Cette division du cadre en deux rectangles égaux révèle les repères sur lesquels le dessinateur met l'accent. Dans le cadre de la sémiotique visuelle, "on peut toujours, par exemple, avoir recours — au niveau de l'analyse du matériau en jeu — à des formes structurées géométriques, qui n'ont pas nécessairement un correspondant sur le plan sémantique" (Courtés, *Du lisible* 247). Le parallèle est évident entre les deux figures. La symétrie est accentuée par les détails graphiques: les deux valises, les masques, les points noirs de pollution. Les deux valises indiquent un séjour prolongé. Plutôt que de brèves vacances, l'intention des deux personnages est de s'installer pour un certain temps dans leurs lieux de destination respectifs. Les masques, au contraire, renvoient aux lieux d'où ils viennent: un masque à gaz et un masque de plongée. L'un est utilisé en ville pour éviter les effets des gaz polluants ou dangereux, l'autre dans les profondeurs sous marines. À ces motifs graphiques renvoyant à des détails de la vie, vient s'ajouter dans le rectangle de droite, un autre détail d'une égale importance. Les diverses tâches sur la peau transmettent le même message que celui du masque à gaz; le lieu de départ du personnage à droite est aussi périlleux que celui de son compatriote à gauche.

Au plan du contenu, l'action suggère un avant et un après. Chacune des deux figures actorielles cumule des rôles figuratifs convergents, en dépit du contraste apparent. Celle du Parisien occupant la moitié gauche de la scène, évoque un départ hâtif, une fuite peut-être? Le costume, le sac à dos, nous rappellent les habitants de la capitale qui se pressent à la sortie du métro en route vers leur travail. Mais apparemment, le Parisien du dessin vient d'y tourner le dos, en route vers une destination différente. La figure de droite est à première vue, en tout point opposée à celle de gauche. Au lieu du costume, il s'agit d'un maillot de bains. La menace qu'illustre le masque à gaz, fait contraste avec l'ambiance de loisirs qu'évoque le masque de plongée. Les chaussures d'une tenue soignée, s'opposent aux tongs. Leur destination représente un autre aspect contradictoire. En dépit des oppositions entre ces divers motifs, une symétrie frappante unit les deux figures. Elles se rapprochent, dans le dessin, par leur taille, leur forme, la part que chacune occupe, les paroles qu'elles échangent, leur ombre et leur mouvement.

Au niveau narratif, la structure du dessin évoque trois étapes. En situation initiale, il s'agit d'un seul thème: le départ. On est déjà en fin d'après-midi, les ombres sont allongées et les deux actants, après avoir longuement attendu, se décident à quitter un lieu où il n'y a pour eux que menace et danger. Le masque à gaz et les tâches sur la peau rappellent ce fait constamment. La deuxième phase de l'action est celle qui se joue sous les yeux du lecteur. La rencontre imprévue des deux actants, les brèves paroles qu'ils échangent, les points d'interrogation successifs dans les deux phylactères, sont suffisamment de preuves de leur étonnement, voire de leur inquiétude. La situation finale est ainsi anticipée. Ils ont déjà deviné intuitivement la triste réalité. Leur projet s'avère irréalisable et chimérique. Pour l'un comme pour l'autre, la fuite est impossible, les vacances dont ils rêvent sont avortées.

Si nous voulons adopter le parcours narratif du schéma actantiel selon Courtés, les deux figures seront réciproquement l'une pour l'autre à la fois, sujet et anti-sujet. La marche du Parisien, actant-sujet, est entravée par la rencontre de l'anti-sujet, le Niçois dont le progrès est à son tour bloqué par le Parisien, devenu anti-sujet. Sans en avoir l'intention, chacun des deux sujets prend l'autre pour objet. Le parcours de la signification reposant sur une opposition sémantique, leur action se traduirait au niveau thématique par les états: réalité vs chimère. Le message de l'énonciateur est alors clairement transmis à l'énonciataire. Il faut faire face à la réalité: la pollution règne partout. À moins qu'une action décisive et rapide ne soit entreprise à tous les échelons, personne ne pourra y échapper.

Dès la première vue, le plan figuratif du dessin suivant (Figure 2) se caractérise par la disproportion entre les diverses formes graphiques. La figure du monde industriel occupe les trois cinquièmes du dessin. En même temps, par sa taille minuscule, le misérable protestataire, cloué sur la fourchette que tient le géant, nous rappelle le combat inégal opposant David à Goliath. Sur le plan horizontal, en divisant le dessin en trois parties égales, le rectangle central regroupe l'ensemble des protestataires dont deux sont encore plus minuscules que leur leader. La variation des motifs attire notre attention par sa signification même. Le globe terrestre figurant dans le rectangle inférieur, près de la jambe du géant, ne paraît avoir qu'une moindre importance. Le couteau et la fourchette brandis entre ses mains, la serviette nouée autour du cou et l'assiette vide, soulèvent tous cette question: le géant a-t-il terminé son repas? Mais tout de suite, la bouche entrouverte suggère une réponse négative.

Par leur taille, par l'expression d'effroi qui se dessine sur leur visage, les deux personnages figurant dans le rectangle central illustrent l'impuissance. Aucune inscription n'est présente sur les affiches qu'ils portent, comme s'ils n'osaient plus s'exprimer. En effet, sur le plan vertical, le misérable protestataire dans le rectangle central devient la victime du géant. Enfin, sur le tiers horizontal supérieur du dessin, aux caractères minutieux mais lisibles que brandit le protestataire, s'oppose en grands caractères, dans le hors-cadre, la menace provocante de la pensée du géant, avec — en réponse à l'affiche de l'activiste — un grand point d'exclamation. Le contexte spatio-temporel est bien défini par la carte du monde et renvoie, par une allusion évidente, à l'avenir de la planète.

Le niveau narratif se joue donc dans le tiers central du dessin, tant sur le plan horizontal que vertical. Le dessin de presse a l'avantage d'une lisibilité immédiate. Il suggère un avant, un pendant et un après. Aussi est-il nécessaire de reconstituer le parcours de la signification. Celui-ci est toujours lié au contexte socioculturel comme à l'actualité évoquée. Après une segmentation du dessin qui s'attarde à étudier — en premier apport — l'énoncé visuel, il est indispensable de prendre en considération l'énoncé verbal. Les cinq affiches des protestataires, ne portent aucun message, sinon celui d'un vide impuissant. S'agit-il des voix des consommateurs, lesquelles ne sont plus écoutées car reléguées au bout du dessin? Il est possible que ces protestataires et leur leader ne soient que les activistes groupés par milliers au dehors des salles de conférence où se tiennent les réunions de l'OMC ou lors du sommet annuel du G8. Contre l'excès du recours aux OGM dans les produits alimentaires commercialisés, leur protestation demeure complètement inefficace. Leur attitude passive n'est d'ailleurs que l'aboutissement des deux scènes qui les précèdent. Une situation initiale où le géant loin d'avoir terminé son repas, se détourne vers le leader et se prépare à le dévorer. Celle-ci est suivie d'une situation intermédiaire où le leader des Verts ose quand même brandir son affiche qui illustre le déséquilibre entre la logique du monde capitaliste et celle du consommateur exploité.

Les personnages légers qui occupent une place minime dans le dessin défient à leur tour le personnage géant qui révèle la logique du monde industriel. Le parcours thématique est celui de: puissance vs faiblesse. Le géant en situation initiale, sûr de lui, affirme sa puissance qui ne tarde pas à être défiée par l'insistance de l'activiste et son groupe, les Verts. Nous passons de la puissance à la non-puissance qui peut être définie, dans notre cas, par un carré sémiotique. Le protestataire audacieux dont la faiblesse est apparente, est bientôt dans une situation de non-faiblesse, ne serait-ce que momentanément, car il continue à brandir son affiche. La réaction de ses collègues cependant, nie cette non-faiblesse qui culminera au niveau de la puissance du géant; point de départ du carré sémiotique.

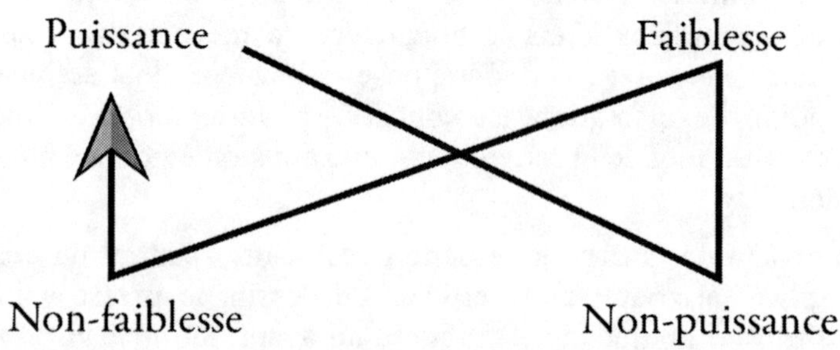

Le dessin suivant (Figure 3) s'attaque aussi aux pays industrialisés, désignés par l'usine et par la fumée. Ceux-ci refusaient en 1997 de réduire leurs émissions de gaz à effet de serre. Face aux accusations du reste du monde, le grand homme du triangle de droite — confortablement assis sur le banc de l'accusé — refuse de reconnaître sa culpabilité. Au contraire, il "propose un arrangement à l'amiable." Le dessin, coupé en deux triangles par une diagonale du bas à gauche vers le haut à droite, place le globe terrestre, motif qui représente l'ensemble des travailleurs exploités de notre planète, en position plus élevée. "Cette disposition topologique exprime la supériorité de l'un des deux antisujets sur l'autre" (Everaert-Desmedt 231). Les autres formants figuratifs dans les deux triangles renvoient au conseil des Prud'hommes. Le couplage des formes contenues dans les deux triangles établit un parallèle et une opposition. Les deux avocats, avec leur toge, les deux bancs où se tiennent d'habitude le salarié et son employeur sont des similitudes entre les triangles. Cependant, par sa taille comme par sa réaction, les quatre gouttelettes de sueur et son étonnement évident, le monde des militants/travailleurs insiste sur la position de force qu'il oppose à l'attitude arrogante de "l'employeur." Le juge ou conseiller occupe sa place dans le triangle des salariés. Sa disposition dans l'image avec sa main gauche levée et pointée vers l'employeur, évoque son incapacité à trancher le litige. Chose étrange car, en général, le conseil des Prud'hommes est une juridiction relativement rapide et efficace.

Sur le plan conceptuel, le parcours narratif est le suivant: disjonction vs conjonction. L'avant est bien connu parce que mentionné dans les titres de l'actualité: refus de la part des grands pays industrialisés d'accorder au reste du monde sa demande de réduire les gaz polluants. Cet état de disjonction est suivi de la non-disjonction soulignée par l'attitude de l'accusé et se transforme en une conjonction possible soulignée par la proposition d'un arrangement. Mais le monde industriel a beau proposer un compromis pour parvenir à un accord avec le reste du monde, le refus obstiné et l'indignation du globe dans le dessin redonne plus de force à la situation initiale de disjonction.

La scène de tribunal qui se joue sous les yeux du lecteur renvoie à un après indécis. Alors que la justice prud'hommale parvient dans la majorité des cas à obtenir une satisfaction partielle ou totale en faveur du salarié, ce procès en particulier semble voué à un échec. Au lieu de faire respecter les textes légaux et conventionnels relatifs au droit des travailleurs, citoyens du monde entier, le juge demeure impuissant. Loin de renier les accusations, les industriels recherchent un compromis. Leurs mensonges sont dévoilés dans le phylactère qui se termine par deux points d'exclamation, soulignant l'indignation et la protestation du monde. Au plan de l'action, le globe, plus grand et placé à un niveau supérieur à celui de l'accusé, continue à être méprisé. Notre monde insiste à réclamer son droit à respirer de l'air pur. En revanche, le businessman continue à émettre la fumée qui remplit toute la partie supérieure du triangle de droite et le dépasse pour occuper une part de celui de gauche. À ce défi flagrant vient s'ajouter la proposition du "patron." Le signifié n'est rien d'autre qu'une interrogation posée par le dessinateur: le monde parviendra-t-il à obtenir son droit le plus légitime, celui de pouvoir respirer, ou devrait-il aller en appel et poursuivre la lutte pour que lui soient accordées ses revendications?

Plantu s'est tourné vers les thèmes écologiques à partir de 1970. Des centaines de ses caricatures s'attaquent aux problèmes de l'environnement. Or, la particularité de ces dessins est qu'ils laissent une large place à l'interprétation du récepteur. Leur fonction pragmatique et leur impact sur les lecteurs font la richesse de cette forme d'expression qui, tout en demeurant au sein même de l'actualité, y ajoute une nuance intéressante. En ce sens, son crayon à la main, Plantu — "maître de la langue de bois" (Payot) — communique un message cohérent et significatif. Dans les dessins qui suivent, on pourra aisément repérer les motifs visuels et les formes graphiques pour passer ensuite au niveau narratif et celui de l'action et attribuer à chacun une interprétation sémantique. En dessinateur engagé, Plantu emploie sa plume et rejoint la lutte pour sauver notre environnement. Ses "figures visuelles" "deviennent des signes-objets" (Courtés, *Du lisible* 230) qui transmettent un message évident. Il nous donne ainsi son interprétation des problèmes d'actualité s'écartant des articles des journaux (Darras 23).

Après la pollution de l'air vient celle de l'océan (Figure 4). Le grand navire américain — dont l'identification est facilitée grâce au motif du drapeau — vu sa dimension, occupe le triangle gauche du dessin. En même temps, sur la large base du triangle droit sont parsemés des déchets industriels dont le navire se débarrasse. En outre, un rapport de contraste dans les intensités lumineuses s'établit entre les deux triangles. Le navire en blanc est entouré de noir et le tiers inférieur est tellement sombre qu'on n'y voit pas la griffe du dessinateur. Au niveau narratif, le pêcheur à la ligne sur le bateau minuscule, la tête levée avec étonnement, raconte un avant prospère et un après menacé de pénurie. Aucune autre figure humaine n'apparaît; car évidemment, les actes perpétrés par le navire ne sont guère humains.

Le dessin suivant (Figure 5) évoque de même l'insouciance des riches de ce monde. Au lieu d'être coupée par une diagonale, cette caricature pourrait être divisée par une médiane horizontale. Le rectangle inférieur évoque une marche hâtive et une poursuite, alors que le rectangle supérieur comprend les figures des acteurs et les bulles révélant par des motifs évidents, l'histoire que raconte le dessin. Les messieurs avec leur costume de politiciens, leur tenue formelle, leur sac et leur pas pressé, symbolisent tout ce que l'Occident développé représente. Leur réaction indignée face à cet être misérable qui essaie en vain de les rejoindre, le regard sérieux, vide de compassion et les deux points d'exclamation qui marquent la fin de leurs paroles, le tout souligne leur insouciance, voire leur mépris du véritable problème. Le tout se réduit selon eux, à la manipulation de "leurs" tiers-mondistes. La figure de gauche est en tout point opposée à celles de droite. Elle évoque un Africain, reconnaissable à la couleur de sa peau comme à la misère de ses vêtements, ses pieds nus et sa taille mince. Il porte un bol vide auquel répond son cri de désespoir accentué par les trois points d'exclamation. Le niveau thématique révèle l'opposition suivante: besoin vs indifférence. C'est l'absurdité plus que la méchanceté que condamne le dessin et le jeune Africain semble plus attristé que révolté. Comme si les citoyens du tiers-monde "nous plaignaient, plus qu'autre chose, de nous désintéresser d'eux, de nous empiffrer pendant qu'ils crèvent de faim" (*Pas nette* 5). En note accompagnant ce dessin, Plantu confirme que, lors de sa parution en 1984, "2 milliards d'êtres humains sont privés d'accès à l'eau potable, et 450 millions souffrent de la faim" (114).

Le dernier dessin (Figure 6) rappelle le premier mais se caractérise par son message plus tranchant. La caricature se divise en un rectangle à droite qui occupe les deux tiers de l'image et un rectangle à gauche où se tient "Monsieur-tout-le-monde" (Everaert-Desmedt 227). Les Parisiens — habitants de la ville — auparavant reconnaissables par les volants, leur tenue formelle et des visages sérieux, ne sont plus reconnus. Le seul motif auquel ils seront dorénavant associés est le masque à gaz. L'air pollué couvrant les deux-tiers de la caricature est repérable par les points noirs à l'arrière fond. Isolés du monde extérieur qu'ils ont contaminé, ces citadins sont condamnés à vivre emprisonnés dans leur costume de protection. Monsieur-tout-le-monde en situation initiale croit avoir reconnu des "clones" identiques. Les Parisiens dont l'étonnement est marqué par le point d'exclamation, dévoilent alors leurs origines. Couverts de la tête aux pieds, les habitants des grandes villes pourront-ils fuir les effets de la contamination de l'environnement? Quel avenir attend notre planète?

En appliquant cette approche sémiotique aux dessins de Plantu, notre objectif était d'en déceler les signifiés. Nous avons retracé le parcours interprétatif suivi par un destinataire qui est au courant de l'actualité. Notre analyse demeure, toutefois, plutôt subjective. Prétendre prendre en compte une matérialité qui serait strictement objective, s'éloigne du sens même du visuel. Naturellement, vu les différents socioculturels, il peut y avoir des divergences d'interprétation par rapport au même objet/dessin. Il y aura toujours cette "autre culture, autre 'grille de lecture' ou interprétation sémantique possible" (Courtés, *Du lisible* 249). Comme tout objet sémiotique, le dessin de presse comporte un plus auquel un œil plus averti pourra parvenir. Bien entendu, d'autres constituants sont exploitables dans chaque caricature et d'autres interprétations sont possibles. L'objectif de notre analyse est de dévoiler, ne serait-ce que partiellement, l'originalité des dessins de Plantu qui fait que leurs unités constituantes se complètent pour produire un signifié unique. Dans notre cas, ils sonnent les glas pour avertir le monde des dangers imminents dans une "planète oubliée."

Figure 1. Plantu, *La France dopée* 151. Dessin utilisé avec l'aimable autorisation de Plantu © 2012

Figure 2. Plantu, *Le petit écologiste illustré* 18. Dessin utilisé avec l'aimable autorisation de Plantu © 2012

Figure 3. Plantu, *La France dopée* 61. Dessin utilisé avec l'aimable autorisation de Plantu © 2012

Figure 4. Plantu, *Pas nette la planète* 121. Dessin utilisé avec l'aimable autorisation de Plantu © 2012

Figure 5. Plantu, *Pas nette la planète* 114. Dessin utilisé avec l'aimable autorisation de Plantu © 2012

Figure 6. Plantu, *Le petit écologiste illustré* 33. Dessin utilisé avec l'aimable autorisation de Plantu © 2012

Ouvrages cités

Courtés, Joseph. *Analyse sémiotique du discours. De l'énoncé à l'énonciation*. Paris: Hachette, 1991.

______. *Du lisible au visible.* Bruxelles: De Bœck-Université, 1995.

Darras, Bernard, et al. *Images et sémiologie. Sémiotique structurale et herméneutique*. Paris : Publications de la Sorbonne, 2008.

Everaert-Desmedt, Nicole. *Sémiotique du récit*. Bruxelles: De Bœck-Université, 2000.

______. "Le Parcours génératif de la signification." *Sémiotica*, Vol. 94 (1-2), 1993: 120-34.

Payot, Marianne. "Plantu, le trait et la politique." *Lire*. Décembre 1995. <http://www.lexpress.fr/culture/livre/plantu-le-trait-et-la-politique_799480.html>

Plantu. *Pas nette la planète*. Paris: La Découverte et Le Monde, 1984.

______. *La France dopée*. Paris: Seuil, 1998.

______. *Le Petit Écologiste illustré*. Paris: Seuil, 2002.

Leon Sachs

University of Kentucky

Eco-Lessons: Yann Arthus-Bertrand's *HOME* and the Republican Pedagogical Contract

This article situates Yann Arthus-Bertrand's 2009 ecofilm, *HOME*, in a pedagogical tradition of active, sensualist learning that, while not exclusively French, has long been central to French republican educational ideology. It specifically examines the film's use of the estrangement effect, a manifestation in aesthetic terms of object lesson pedagogy, in order to provoke a more careful scrutiny of its principle object of study, the Earth.

Like Al Gore's *An Inconvenient Truth* (2006) and Leonardo DiCaprio's *11th Hour* (2007), Yann Arthus-Bertrand's *HOME* (2009) is a didactic film. Its message is clear. It emphatically calls attention to the problem of climate change, the rapid depletion of the Earth's resources and other harmful effects of modern industrial production. *HOME*, like these other films, is also a call to action: it entreats viewers to become more conscientious and responsible stewards of the planet, to become what is routinely termed global citizens. And yet, the film delivers this now familiar message in remarkably unfamiliar ways.

HOME relates the story of the Earth's origins and stages of development up through man's discovery of oil and the environmental havoc that this discovery would eventually wreak. Its stunning aerial images of the Earth taken from 120 different locations in 54 different countries document this dramatic saga with a vividness that no ordinary reportage could ever achieve. The esthetic and narratological innovations in this cinematic montage distinguish it from the more traditional modes of transmitting information that we find in the films of Gore and DiCaprio.

This essay examines more explicitly how Yann Arthus-Bertrand and (screenwriter) Isabelle Delannoy's original esthetic serves a pedagogical and, ultimately, political purpose.[1] In fact, I will argue that Arthus-Bertrand's esthetico-pedagogical technique is not as original as it might at first seem. His bewildering images of the earth's surface arouse the viewer's faculties of perception in a manner that recalls a republican pedagogical tradition emphasizing direct, sensory observation of concrete objects. This pedagogy, known as intuitive or object lesson pedagogy, was not only an elementary form of scientific inquiry. It was also a mechanism for teaching republican values rooted in Enlightenment rationalism; it gave concrete form to abstract notions of citizenry and civic duty. In other words, a pedagogy of scientific observation also promoted the spread of republican ideology and thereby fostered the emergence of a national community united around a common set of civic values. *HOME*'s global citizen is, in many respects, a republican citizen on a planetary scale.

The film's distribution indicates already an attachment to the republican educational tradition. It was released worldwide on June 5, 2009 on World Environment day and made available in all media formats — movie theater, television, DVD and internet — free of charge (DVDs are sold at cost). Translated into nearly 40 languages, the film has been viewed by more than 200 million people in 131 different countries; an estimated 14 million people have viewed it on YouTube.[2] The large-scale, nonprofit distribution of the film signals the filmmaker's commitment to universal access to knowledge. While this may not be an exclusively French preoccupation, it is certainly a cornerstone of the French republican tradition inherited from the Enlightenment, most famously typified by Diderot and d'Alembert's Encyclopedia.

---

[1] Space does not permit a listing of all those who, along with Arthus-Bertrand and Delannoy, contributed to the writing of the scenario. As is customary in discussions of film, I will refer to *HOME* as principally the work of its credited director, Yann Arthus-Bertrand.

[2] This information was communicated in a press release received from Yann Arthus-Bertrand's foundation, Goodplanet.

The film's ties to this tradition become more pronounced when we consider that it is accompanied by a *dossier pédagogique*, a free, online instructor's manual (available in six languages) intended to help teachers incorporate the film into their lessons. The instructor's manual, moreover, was created by the *Ligue de l'enseignement*, a one hundred and forty-year-old organization dedicated, in the republican spirit, to the spread of learning. Founded in 1866 by Jean Macé, who, like Arthus-Bertrand, was a popularizer of scientific learning, the *Ligue* played a leading role in the establishment of free, mandatory and secular education for all French children under the Third Republic.[3] The film's republican genealogy, in other words, can be traced through the founding of the *école Ferry* at the end of the nineteenth century.

The more interesting republican features of the film, however, are embedded in its form, in the visual and linguistic devices and narratological structures that stimulate the viewer's observation. These reveal themselves already in the film's prologue and very first image (Figure 1). When the word *HOME* appears in capital letters beneath a slowly-turning, partially-obscured picture of the planet Earth, the viewer understands immediately what is being communicated. There is an effort to create a sense of community. The combination of word and image asks the viewer to imagine membership in a community so vast that, like Benedict Anderson's nation, it eludes direct and complete observation. The opening segment attempts to create a sense of proximity to — and intimacy with — that which lies beyond any kind of immediate reach. To use the language of object lesson pedagogy, we could say that this image — like many of the images in the film and the film as a whole — tries to make familiar and concrete that which can too often seem far away, foreign and removed from daily, lived experience.

This effort to reduce the distance between the near and the far continues in the spoken words of the film's prologue. Still with the opening image of the planet on the screen, a voice-over says the following in a calm, measured tone:

> Ecoute-moi s'il te plaît.

---

[3] On Jean Macé and the history of the *Ligue*, see Katherine Auspitz, *The Radical Bourgeoisie: The* Ligue de l'enseignement *and the Origins of the Third Republic, 1866-1885.*

> Tu es comme moi un “homo sapiens,” “l’homme qui pense.”
>
> La vie, ce miracle dans l’Univers, est arrivée il y a presque 4 milliards d’années.
>
> Et nous les Hommes il y a seulement 200.000 ans.
>
> Et pourtant, nous avons réussi à bouleverser cet équilibre.
>
> Écoute bien cette histoire extraordinaire qui est la tienne et décide alors de ce que tu veux en faire.

With these words, a kind of *captatio benevolentiae*, the narrator addresses the listener in a familiar manner. A special relationship is established between speaker and listener, both members of the same rational species. It is as if this avuncular voice were calling his kin to gather around — as if by his feet, around the hearth at home — to listen to the telling of an “extraordinary,” “miraculous” tale. The conventional “once upon a time,” however, is replaced by a startling 4 billion years ago. Humankind is but a mere 200,000 years old. Its history is but a recent, brief chapter appended to a much longer, almost unfathomably longer, story. The introduction to the narrative performs a function similar to that which we observed in the juxtaposition of the planet and the film’s title. At the same time that it reminds us of the awe-inspiring temporal gulf separating the present moment from the primeval age, it closes that gap by insisting that this much longer story is also man’s own — that it is still with us today.

A brief comparison with George Bruno’s school manual of 1877, *Le Tour de la France par deux enfants* — especially with the manual’s preface — serves to put in still greater relief the film’s debt to a republican pedagogical tradition. The best-selling primer and “site of republican memory” tells the story of André and Julien Volden, two schoolchildren who leave their home in German-occupied Lorraine after the Franco-Prussian War in order to return to France.[4] Their journey leads them through different French regions where they learn firsthand about the nation’s history, customs, geography, industries and products. The manual tried to educate its young readers about their membership in — and responsibility toward — a much larger national community. As

[4] See Jacques Ozouf and Mona Ozouf, “*Le Tour de la France par deux enfants*. Le petit livre rouge de la République.”

Bruno's preface makes clear, the story attempted to give concrete form to the abstract concept of *la patrie*:

> La patrie ne représente pour l'écolier qu'une chose abstraite, à laquelle, plus souvent qu'on ne croit, il peut rester étranger pendant une assez longue période de la vie. *Pour frapper son esprit, il faut lui rendre la patrie visible et vivante.* Dans ce but, nous avons essayé de mettre à profit l'intérêt que les enfants portent aux récits de voyage. En leur racontant le voyage courageux de deux jeunes Lorrains à travers la France entière, nous avons voulu la leur faire pour ainsi dire *voir et toucher*. (Bruno iv, my emphasis)

As we have seen, Yann Arthus-Bertrand's film also uses the device of storytelling to enliven — to bring closer to home, so to speak — man's relationship to the Earth. Though there are no individual human characters in *HOME* as there are in Bruno's story, its protagonists — Earth, Man and oil — are actors in an epic story about the rise and fall of man's harmonious relationship to his environment. Like *Le Tour de la France*, the film attempts to instill a sense of "civic" duty with respect to a much larger — in this case, planetary — community.

We also hear in Bruno's preface the language (in italics above) of sense perception, of seeing and touching. This word choice indicates Bruno's debt to the quasi-official republican pedagogy of the object lesson and intuitive learning. Derived from the sensualist, empiricist theories of Locke and Rousseau (among others), this pedagogy, as I mentioned above, insisted on the importance of studying concrete things before abstract concepts, on sensuous — especially visual — learning over bookish learning and words. It emphasized the student's active involvement in the learning process and disparaged the passive reception of information. Often knowledge of the specific object of observation was considered less important than was the cultivation of the student's skills of observation and analysis in and of themselves. Great emphasis was placed on awakening the faculties of observation so that perception should not become dull, passive and routinized.

In *Le Tour de la France*, the adventures of André and Julien enact this experiential, sensualist education. Their education depends on their direct encounters with people, places and things. Little Julien, for example, learns about the Cantal mountains not because he has read about them but because he has climbed them with his own two feet. He understands the engineering genius of the Marquis de Vauban because he has seen Vauban's fortress walls with his own two eyes. Thanks to the manual's more than 200 realist engravings, the reader also studies these places and things in precise visual detail.

Yann Arthus-Bertrand applies, it would seem, a similar pedagogical doctrine. He intends to have us see and touch the planet. He approximates this effect not only through the device of storytelling but also, and more strikingly, through the use of visual imagery. One of Yann Arthus-Bertrand's signature talents as a photographer is his ability to throw into detailed relief, give texture to, and render almost palpable the things and places he puts before our eyes. His use of framing, shot angle, panning and zoom techniques does more than merely show us what he sees. His art awakens our own faculties of observation. It requires that we look actively, closely and carefully. Again, the language of republican educators captures the nature of this learning experience. As Ferdinand Buisson, Jules Ferry's leading lieutenant, wrote in 1911:

> Si l'exercice aiguise, affine et fortifie nos facultés de perception externe, ce ne sont pas seulement les sens qui y gagnent, c'est le jugement: l'intuition sensible alors devient l'observation; l'enfant n'est plus seulement capable de voir ou d'entendre, il sait regarder et écouter; il ne subit plus des sensations, il sait les faire naître, les diriger, les comparer, les distinguer, les mesurer, les analyser; il observe. ("Intuition")

This pedagogy turns seeing into looking, hearing into listening. It transforms passive reception of sensory information into active observation, judgment and analysis. The exercise is coterminous with rational inquiry. Arthus-Bertrand's images implement this pedagogy. We do not only *see* them, we *look* at them in terms analogous to the concept of observation at the heart of object lesson pedagogy.

It is here, however, that the paradox begins to emerge. The power of Yann Arthus-Bertrand's art depends in large part on an esthetics of estrangement. In order to have us look closely at the earth's surface, at those places where natural beauty and environmental destruction coincide, the filmmaker presents his subject as strange and unfamiliar. Often, Arthus-Bertrand has us look at an image (accompanied by a gradual zoom or changing camera angle) without letting us know exactly what we are observing. True to the object lesson tradition, images take precedence over words. While many of the images in the film illustrate this point, it is particularly apparent in the view of the Pjorsa River in Iceland (Figure 2). It is a sensuous image, the detailed almost baroque patterns produce a highly-textured, almost palpable effect. Even once the voice-over tells us that this is water, the image continues to give us pause. It is unlike any view of water we have had before. It is a strange, almost unrecognizable image, and it is its strangeness that forces us to examine it more closely. It is paradoxically through estrangement and defamiliarization that the artist brings the planet closer to home.

Space does not permit a full review of the theory of artistic estrangement.[5] Suffice it to say that this staple of twentieth-century esthetic theory and formalist criticism is concerned above all with the artistic devices that awaken the reader's or spectator's otherwise dormant, dulled or routinized faculties of perception. To paraphrase the Russian formalist, Victor Shklovsky, whose name has become synonymous with the concept, the purpose of formal technique in art is to make "objects unfamiliar" so as to make the activity of perception more rigorous (Shklovsky 778). Once considered a concept serving only apolitical and ahistorical questions of artistic form, it is now viewed as a useful tool of both politically-committed art and the analysis thereof — as evidenced, of course, by Yann Arthus-Bertrand's film.

What is more, the concept of estrangement, though not recognized as such, plays a role in republican pedagogy with respect to the imperative of stimulating the faculties of perception. In *Le Tour de la France par deux enfants*, little Julien expresses constant bemusement before the curiosities he encounters and only comes to understand them after careful, more measured, scrutiny. In her study of the influence of republican education on modernist visual arts, Molly Nesbit explains how the most commonsensical exercise of object lesson pedagogy could lead to analyses and discoveries that defied common sense. Emile Durkheim, in his Sorbonne lectures on the evolution of pedagogy in France, provides what is perhaps the most powerful expression of estrangement as applied to republican education. Durkheim argued that the modern,

[5] See Carlo Ginzburg, "Making Things Strange: The Prehistory of a Literary Device."

"scientific" study of literature should foster an examination of man and humanity not as an already understood model of perfection (as was the custom in the classical humanities), but rather as an ever-changing "thing" ("une chose humaine") and an "unknown reality" ("une réalité ignorée"). Such studies would have the beneficial effect, he said, of disorienting ("dépaysement") the student (Durkheim 367-82).

There is almost an echo of Durkheim's point in the film's use of the term *homo sapiens* in the prologue. As I remarked earlier, these words are uttered by the narrator to establish an identification and trusting relationship with the viewer/listener. They establish a sense of intimacy and contribute to the goal of making this distant story seem close to home. Nevertheless, the Latin moniker also produces the opposite effect. The language of biological taxonomy identifies man as just another species among millions to be observed with a dispassionate, objective eye. He becomes, to borrow Durkheim's words, "une chose humaine" and "une réalité ignorée." The prologue, in other words, establishes in the same breath a sense of closeness to — and estrangement from — the story about to be told. It is a paradoxical mode of teaching and learning that characterizes the film as a whole.

In conclusion, I would like to stress the value of situating *HOME* in a history of republican education and pedagogy. We should first observe that drawing attention to the question of hidden, foreign and unfamiliar history is a central concern of the film itself. One of its principal claims is that mankind's history must be understood as part of Earth's history, the present as part of the past. Responsible global citizenship depends on knowledge of this historical relationship. It thus follows that examining the way an educational film about climate change belongs to a longer history of education — in particular, a brand of education committed to promoting collective identity and civic duty — is consistent with a basic premise of the film's historical outlook. To put it another way: ecological sensibility calls for historical awareness.

But an awareness of this history of education must also shape the way we interpret the film's message, especially the way we understand the concept of *home* itself. For, as we have seen, object lesson pedagogy is as much about developing the faculties of perception as it is about gaining knowledge of a particular object. When we explore this history, we discover that successful stimulation of such active, critical observation benefits from artistic devices and learning methods that render the object strange and unfamiliar. We encounter this both in the history of sensualist pedagogy and in Yann Arthus-Bertrand's pedagogical film. This leads, however, to unexpected conclusions about what would appear to be the overarching thesis of the film: that we must all understand the planet as our collective home, our closest community and thus protect it accordingly. If this act of understanding depends on careful observation and if such observation benefits from the experience of estrangement — clearly a hallmark of Yann Arthus-Bertrand's esthetic — then perhaps a hidden message in the film is the opposite of what its title first suggests. Perhaps in order to perceive, know and protect our collective, planetary home, it is necessary that we not feel *too* at home.

Figure 1. Still image from the film's opening sequence.

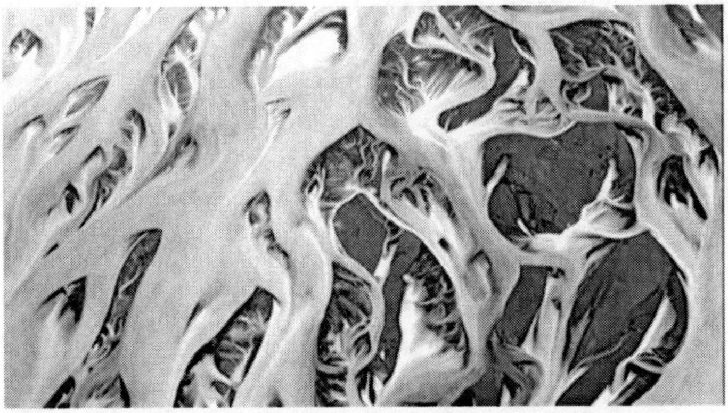

Figure 2. Still image of the Pjorsa River in Iceland.

Works Cited

Anderson, Benedict R. *Imagined Communities: Reflections on the Origin and Spread of Nationalism*. Rev. and extended ed. London; New York: Verso, 1991.

Auspitz, Katherine. *The Radical Bourgeoisie: The* Ligue de l'enseignement *and the Origins of the Third Republic, 1866-1885*. Cambridge: Cambridge University Press, 1982.

Bruno, G. [Augustine Fouillée]. *Le Tour de la France par deux enfants*. [1877]. Paris: E. Belin, 1976.

Buisson, Ferdinand. "Intuition." *Nouveau Dictionnaire de pédagogie et d'instruction primaire*. Vol. 1. Paris: Hachette, 1911. <http://www.inrp.fr/edition-electronique/lodel/dictionnaire-ferdinand-buisson/>

Durkheim, Émile. *L'Évolution pédagogique en France*. Paris: Presses Universitaires de France, 1969.

Ginzburg, Carlo. "Making Things Strange: The Prehistory of a Literary Device." *Representations* 56 (1996): 8-28.

Nesbit, Molly. *Their Common Sense*. London: Black Dog Publishing, 2000.

Ozouf, Jacques and Mona Ozouf. "*Le Tour de la France par deux enfants*. Le petit livre rouge de la République." *Les Lieux de mémoire*. Ed. Pierre Nora. Vol. 1. Paris: Gallimard, 1997. 277-301.

Shklovsky, Victor. "Art as Technique." *The Critical Tradition: Classic Texts and Contemporary Trends*. 1917. Ed. David H. Richter. 3rd ed. Boston: Bedford/St. Martin's, 2007. 774-84.

Isabelle Delannoy

Ingénieur agronome spécialisée dans les questions écologiques, écrivain, co-auteur du film *HOME*, réalisé par Yann Arthus-Bertrand, 2009, Présidente co-fondatrice de l'agence d'intelligence écologique Good Save the Green.

Sciences, humanités et écologie — trouver les nouvelles passerelles et les nouveaux mots pour entreprendre un nouveau monde

L'ampleur de la crise écologique actuelle interroge l'humanité dans son entièreté. Sa dimension implique un changement de vision de nos systèmes techniques mais aussi économiques, politiques et culturels. Elle implique donc un changement de paradigme d'ordre civilisationnel. Parce qu'il ne peut advenir que collectivement, le changement doit s'approprier un nouveau langage capable de projeter tout un chacun dans un avenir par nature inconnu et de rendre visible l'invisible. Sa nature multidimensionnelle implique que l'ensemble des disciplines de réflexion et d'expressions humaines, autant celles dites des "sciences dures" de la matière et de la vie que celles dites des "humanités," dialoguent et collaborent ensemble dans toute leur complémentarité et leur altérité.

Depuis 1950, l'activité humaine a modifié les écosystèmes plus rapidement que pendant toute l'histoire de l'humanité. Les sols voient leur fertilité décroître sur plus du quart des terres émergées, la composition de l'atmosphère change et la température globale de l'atmosphère augmente, les équilibres océaniques se modifient, le rythme d'extinction des espèces est 1000 à 10 000 fois supérieur au taux naturel d'extinction enregistré dans les couches géologiques de la planète. Tandis que les scientifiques prévoient des effets de seuil, qui si on sait qu'ils existent et sont proches, ne sont pas exactement connus. Il n'est pas exagéré de dire que les problèmes écologiques confrontent l'humanité à des enjeux tout à fait nouveaux dans son histoire. L'être humain est devenu la première force géologique du globe et sa puissance technologique de transformation de la matière en est notamment à l'origine. La question posée est donc celle du changement de nos modèles. Ces modèles seraient à tort imputés à l'humanité entière. Ils sont ceux de la civilisation occidentale dont les valeurs, la vision du monde et les techniques ont largement diffusé à travers le monde depuis l'époque des colonisations, depuis celle de la révolution industrielle, ce phénomène s'étant accéléré depuis notamment les années 1950.

Cette diffusion pose néanmoins des questions à l'humanité entière et à tous les peuples et civilisations qui la composent puisque ses conséquences se répercutent à l'échelle du globe: l'atmosphère qui entoure la terre ne connaît pas les frontières humaines, ni la circulation des eaux, ni celle des espèces et des éléments chimiques qui font la stabilité globale de la biosphère. Ils la confrontent à des questions inédites: celle d'une solidarité globale, dans l'espace, mais aussi dans le temps puisque c'est la capacité de subsistance même des générations à venir qui est mise en question. Enfin cette solidarité dépasse même l'échelle humaine. A mesure que les écosystèmes s'épuisent et les espèces vivantes décroissent, l'homme réalise qu'il en est dépendant et que sa survie dépend de la survie des espèces avec qui il cohabite sur le globe.

Il est ainsi commun d'entendre que nous devons "changer de paradigme," c'est-à-dire changer notre "représentation du monde," notre "manière de voir les choses." À ce titre, le changement de paradigme que doivent opérer nos sociétés et chaque individu qui la compose, interroge en premier lieu le langage. Car quel outil exprime mieux notre vision du monde que notre langage qu'il soit celui des mots, des images ou des sons? Si nous devons changer de paradigme, alors nous devons trouver les mots, les champs lexicaux et les formes d'expression qui nous permettront de l'exprimer et plus encore de produire une vision commune de ce que peut être notre avenir, et son appropriation. Changer de paradigme collectivement signifie de fêler nos certitudes, pour rendre le changement de point de vue possible. Comment le faire et quelles techniques utiliser? En outre de quel paradigme parle-t-on? Si l'expression s'est diffusée dans les milieux écologistes ces dernières années à s'en approcher du lieu commun, elle est néanmoins peu approchée dans le détail.[1] Parlons-nous de changement de paradigme économique, technique, ou de civilisation?

Les sciences de la matière et de la vie — ce qu'on appelle communément "les sciences dures" — ont joué un rôle considérable dans l'avènement de cette époque et la prise de conscience de la nécessité écologique: elles ont à la fois produit les techniques de notre puissance et établi le constat et les causes physiques et biologiques de notre vulnérabilité. Elles ont aussi modelé les méthodes et les visions des autres disciplines, ces "humanités" maintenant appelées sciences humaines. Aujourd'hui, quelle est leur place dans la proposition d'un nouveau monde?

J'essaierai ainsi d'argumenter dans les lignes qui viennent en quoi le changement de paradigme qui s'impose est de nature civilisationnelle, pourquoi l'appel aux humanités dans toute leur altérité par rapport aux sciences dites "dures" est primordial et urgent. Je terminerai enfin sur l'importance des disciplines d'expression artistique et littéraire pour la mise en oeuvre de ce changement au niveau collectif.

De quel changement de paradigme parle-t-on? Le changement de nos techniques est-il suffisant pour basculer dans un développement durable de nos sociétés? Dans les grands problèmes écologiques auxquels nous devons faire face dont j'ai cités quelques grands exemples dans mon introduction, figure le dérèglement climatique.[2] Il s'agit de l'un des plus médiatisés. Si la résolution de celui-ci n'assurerait pas à lui seul un avenir durable de nos sociétés, il est un bon exemple pour comprendre que le changement de paradigme nécessaire dépasse le niveau technique et requiert des modifications plus complexes et profondes du

[1] J'appellerai "écologiste" dans cet article, toute personne qui estime, non pas que l'écologie prime en toute chose, mais qu'elle est une dimension incontournable et prioritaire à prendre en compte dans ce début du XXIe siècle, pour assurer l'harmonie de nos sociétés dans le présent et dans l'avenir.

[2] Cette liste n'est pas exhaustive. Y manquent notamment l'introduction humaine de molécules de synthèse — dont on évalue le nombre à 100 000 environ — dans l'environnement qui perturbe les grands équilibres internes du monde vivant, comme la reproduction cellulaire, le système immunitaire, le système nerveux, le système hormonal et reproducteur; l'accumulation de la radioactivité à des taux supérieurs aux taux naturels; les perturbations psychiques liées entre autre chose à l'accélération de la modification de l'environnement immédiat; etc.

fonctionnement de nos sociétés. Le groupe international d'experts pour le climat (GIEC) lui attribue comme cause "très probable" (probabilité supérieure à 90%), l'activité humaine et notamment sa consommation d'énergies fossiles, charbon, pétrole et gaz, émettrice de C02, l'un des principaux gaz à effet de serre (Pachauri 5). Ces énergies fournissent 86% de l'énergie primaire mondiale et la communauté politique mondiale s'accorde sur la nécessité d'une diminution de 80% à l'échelle de 2050, avec un premier pas de diminution de 25 à 40% avant 2020-2035. Notons que face à l'enjeu climatique, ceci est une réponse ad minimum. En effet, les experts climatiques évaluent la probabilité d'éviter la hausse de la température de l'atmosphère mondiale à 2°C — seuil au delà duquel un emballement de la hausse des températures est très probable — de 50% (une chance sur deux) (Guérin et Colombier). La réponse apportée est souvent une décarbonation de l'économie: grâce notamment à une meilleure efficience énergétique de nos économies, c'est-à-dire une diminution de l'énergie consommée par dollar échangé, et au basculement des énergies fossiles vers les énergies renouvelables, nous irions vers une économie et une production "propre." En réalité, nous avons déjà technologiquement entamé les débuts de cette révolution propre: les intensités en carbone ont déjà baissé en moyenne de 0,7 % par an depuis 1990 (Jackson 87). Mais cela n'a pas empêché les émissions d'augmenter de 40% (87-88). Car parallèlement les échanges et la production se sont accrus. Dans son dernier ouvrage, reconnu comme majeur, *Prospérité sans croissance* (2010), Tim Jackson tord définitivement le cou à cette croyance en réalisant quelques calculs arithmétiques simples (77-94). En tenant compte de cette diminution projetée des émissions et de la croissance de la population mondiale attendue à 9 milliards de personnes en 2050, "il faudrait que le contenu moyen en carbone de la production économique soit inférieur à 40 g de CO2/$, soit 21 fois moins que la moyenne mondiale actuelle" (89). En incorporant les objectifs d'équité et de croissance des pays en développement et pauvre, ce facteur s'élève à 55. En prenant en compte les objectifs de croissance communs des pays industrialisés, le facteur s'élève encore à 130! Les seuls changements de technologie ne suffiront pas à atteindre cet objectif, qui on l'a vu déjà est pourtant de très faible ambition par rapport au risque climatique réel. Ainsi Tim Jackson insiste: le changement de paradigme ne peut être que technique, il devra être également économique.

> L'efficacité dans l'utilisation des ressources, les énergies renouvelables et la réduction des flux de matière ont toute un rôle fondamental à jouer pour assurer la durabilité de l'activité économique. Mais l'analyse présentée [...] tend à démontrer qu'il est pour le moins fantaisiste de tabler sur une baisse "considérable" des émissions et de l'utilisation des ressources sans se confronter à la structure des économies de marché. (94)

Autant dire que le changement de paradigme est aussi de nature politique, tant l'économie de marché façonne aujourd'hui la "gestion de la cité."

Qu'en est-il sur le plan intellectuel et des valeurs morales de l'Occident: notre pensée et nos valeurs d'aujourd'hui sont elles compatibles avec les changements requis ou font-elles aussi partie d'un paradigme à changer? Intéressons-nous à une donnée fondamentale dans les enjeux écologiques actuels: les rapports intellectuels qu'entretiennent nos sociétés à la biosphère. L'homme occidental parle communément de "nature" qu'il oppose à l'être humain. Dès la Genèse l'homme est en effet séparé du reste de la Création: lui seul est créé à l'image de Dieu, "Dieu créa l'homme à son image, il le créa à l'image de Dieu" (Genèse 1:27). Il en est en outre symboliquement fait maître en recevant le pouvoir et le droit de nommer chaque être de la Création,

> L'Eternel Dieu façonna à partir de la terre tous les animaux sauvages et tous les oiseaux du ciel, puis il les fit venir vers l'homme pour voir comment il les appellerait. Il voulait que tout être vivant porte le nom que l'homme lui donnerait. (Genèse, 2 :19).

Dans l'Occident, cette dualité Homme-Nature perdure pendant les siècles, dans une conception de plus en plus utilitaire de la Nature, à mesure que la puissance technique de l'homme occidental s'affirme. Cette approche est particulièrement conceptualisée lors de l'époque de la Renaissance comme le montre André Beauchamp dans son *Introduction à l'éthique de l'environnement*: "Bref, sans la dualité homme-nature, la Renaissance aurait affirmé l'homme et réduit la nature à l'état de moyen" (29). A. Beauchamp cite particulièrement l'Anglais Francis Bacon (1561-1626) et le Français René Descartes (1596-1650) comme auteurs "les plus dénoncés dans ce passage à une notion utilitaire de la nature" (29) et par ailleurs parmi les plus influents de la pensée scientifique occidentale. Chez ces auteurs, la connaissance de la Nature est un instrument essentiel du développement de la puissance de l'homme, dont le premier objectif est de la soumettre.

> Pour Bacon, l'intelligence humaine abandonnée à elle-même est impuissante; le secours de la méthode est absolument nécessaire. La seule qui puisse nous conduire à des résultats positifs, c'est la méthode expérimentale. Ce que l'homme ajoute à sa connaissance, il l'ajoute à sa puissance. Le seul moyen de soumettre la nature, c'est d'apprendre à la connaître (29-30),

analyse A. Beauchamp. Dans son discours de la méthode (1637), René Descartes, s'il ne partage pas les mêmes concepts que son prédécesseur n'en partage pas moins la même vision duale du monde. Pour Descartes, continue A. Beauchamp,

> Le doute s'arrête donc à la prise de conscience de soi comme sujet connaissant. Cette expérience permet de reconstruire la certitude. A partir de cette vérité première, Descartes pose l'existence de Dieu et l'immortalité de l'âme, établissant un point de rupture entre l'animal, qu'il décrit comme une machine, et l'être humain animé d'une âme (5e partie). Il estime avoir trouvé la philosophie qui permet de "nous rendre comme maîtres et possesseurs de la nature" (6e partie). (31)

La théorie de l'évolution initiée par Charles Darwin (1809-1892), viendra renverser deux siècles plus tard cette dualité en établissant un continuum entre les espèces et une parenté entre l'homme et le reste du Vivant. Mais 150 ans plus tard, notre vision de la nature dans nos systèmes de production, et d'échange qui font la base de notre économie et de notre vie quotidienne a-t-elle réellement changé? Nous parlons bien davantage de "ressources naturelles" que d'un Vivant à part entière dont l'équilibre et la richesse repose sur la complexité des liens qu'entretiennent les espèces entre elles et avec les éléments minéraux qui les environnent. Nos connaissances scientifiques ont beau chaque jour affirmer et confirmer cet équilibre holistique du vivant, nos esprits n'en ont pas encore pris la mesure. L'agriculteur d'aujourd'hui découvre ainsi avec stupeur que les pesticides qu'il épand sur ses champs pour en éliminer les ravageurs ont sur son système nerveux des effets comparables à ceux qu'ils entretiennent sur les insectes qu'il souhaite éliminer. Nous découvrons que nos chairs accumulent les molécules que nous avons créées, qu'elles perturbent nos équilibres immunitaires, nerveux, reproducteurs ou cellulaires dans des mécanismes comparables aux anomalies que l'on voit se développer dans le monde animal. Ainsi, la nature et l'homme ne sont séparés que dans notre langage. Nos manipulations nous apprennent au contraire et à nos dépens qu'ils sont un même Vivant.

Si l'on considère une civilisation comme l'ensemble des traits qui caractérisent une société tant sur le plan technique, qu'intellectuel, que politique et moral, alors le changement qu'il nous faut opérer dans notre vision du monde est bien un changement de paradigme civilisationnel.

Dans cette entreprise, il semble ainsi illusoire de croire que seules les sciences dites "dures" et leur expression technique puissent apporter des réponses à la hauteur des enjeux. Pour entreprendre la métamorphose de nos civilisations, c'est aussi l'ensemble des humanités, dans toute leur altérité mais en lien et dialogue avec les sciences de la matière, qui semblent devoir entrer en scène à tous les niveaux de structuration de nos sociétés afin d'imaginer et de définir de nouveaux systèmes de production, de consommation, de nouveaux moteurs de reconnaissance sociale et de nouvelle façon de faire lien entre les hommes et avec le reste du Vivant. C'est une réelle métamorphose de nos pratiques et de nos méthodes de pensée et d'action qui est requise. Comme le dit Edgar Morin, sociologue, philosophe et penseur de la complexité, dans son livre *La Voie. Pour l'avenir de l'humanité*:

> Les réformes politiques seules, les réformes économiques seules, les réformes de vie seules ont été, sont et seront condamnées à l'insuffisance et à l'échec. Chaque réforme ne peut progresser que si progressent les autres. Les voies réformatrices sont corrélatives, interactives, interdépendantes. (41)

Ainsi de l'économie par exemple. Si nous considérons l'économie dans sa définition première de "gestion de la maison" (du grec οἶκος, maison et νομοθετῶ, gérer, administrer), et si nous y associons la vision intellectuelle des sociétés occidentales de leur rapport au monde, comme mise en évidence plus haut, d'une nature ressource, à connaître pour "soumettre" et "maîtriser," il semble intéressant de questionner cette économie basée sur notre vision duale homme/nature, sa responsabilité dans la crise actuelle, et de la comparer avec d'autres systèmes économiques mis en place par l'homme dans le monde et à travers l'histoire. A cette question, n'est-ce pas les compétences de l'anthropologie et de l'histoire des sociétés humaines qui apporteront des éléments de réponse? De la même façon, les moyens de faire évoluer nos sociétés, de comprendre ses réticences et d'identifier ses moteurs et ses désirs, ne sont-ils pas de la compétence des sociologues, des psychologues, des historiens?

Les outils de notre reconstruction sont chez les urbanistes les architectes, chez les physiciens, les biologistes, mais aussi chez les sociologues, les psychologues encore, et les historiens de l'art. À quoi nous servira-t-il de construire des "éco-quartiers" de la plus haute performance énergétique et de la meilleure "biodégradabilité" aujourd'hui technologiquement possible, s'ils n'ont pas d'âme? S'ils reconduisent dans leur structures et leurs schémas, les ensembles construits comme on l'a vu en Europe après-guerre à la va-vite, et nid aujourd'hui d'exclusion et de violence sociale? À quoi nous servira encore de remplacer la voiture à combustion par la voiture électrique si c'est pour nous pousser à la consommation d'espaces toujours plus grands et continuer ainsi à fractionner l'habitat des espèces dont nous dépendons pour notre survie, à imperméabiliser des sols dont nous dépendons pour notre nourriture et notre adaptation aux aléas climatiques, à accroître sans cesse la distance entre nos lieux de vie et nos lieux de travail au détriment du temps passé à éduquer nos enfants? À quoi servira-t-il enfin d'améliorer l'efficience en ressource de nos objets de consommation si nous en avons besoin de toujours plus pour affirmer notre reconnaissance sociale? Est-ce sur l'objet qu'il faut travailler ou sur sa finalité?

Ce ne sont que des exemples mais ils montrent l'étendue des changements qu'il s'agit d'opérer dans cette construction d'une nouvelle vision du monde, de nos vies, du sens de nos actions, et de nos métiers. Nouvelle vision ne signifie-t-elle pas en outre nouveau langage, voire nouveaux mots? Allons faire un tour du côté des géologues, ces spécialistes de l'histoire de la planète pour commencer l'exploration de cette question. Claude Lorius est glaciologue et pionnier des recherches sur le climat. Dans son livre *Voyage dans l'anthropocène*, il se fait l'écho des nombreux géologues qui aujourd'hui demandent l'identification d'une nouvelle période géologique, celle de l'anthropocène, pour rendre compte des modifications apportées à la planète Terre de la part de l'activité humaine. Ces dernières sont en effet désormais lisibles dans les strates géologiques qui sont la mémoire de son histoire. Il y écrit:

> Le jour où l'on change de regard, il faut changer le vocabulaire. Le jour où l'on change de monde, il faut changer les noms. Nous ne pouvons plus prendre les choses telles quelles, dans leur simple continuum… Puisque rupture il y a, il faut la nommer pour la voir, pour l'expliquer, pour l'autopsier, voire pour la conjurer. (12)

C'est en effet par les mots, leur choix et leur agencement que nous exprimons notre vision du monde. Dans cette immense entreprise de changement de notre regard, le travail sur notre langage semble une étape parmi les premières d'entre toutes. Pour continuer cet argumentaire, je parlerai notamment dans cette partie de mon expérience en tant qu'écrivain sur les question écologiques et cherchant à en vulgariser les enjeux. Je m'appuierai en particulier sur l'expérience du film *HOME*.[3]

Revenons par exemple sur le terme "nature," auquel nous nous sommes intéressés plus haut. Par son histoire, comme nous avons pu le montrer, il porte en lui le sens profond d'une séparation de l'homme avec la biosphère, que nous savons désormais illusoire. Il fait ainsi partie des mots qui appartiennent profondément à ce paradigme que nous souhaitons quitter. Mieux, il en est peut-être un des marqueurs. Dans le film *HOME*, le mot "nature" et ses dérivés comme "naturel" ont été réservés aux descriptions de notre monde existant et par là, de notre pensée existante. Pour comprendre notre lien à la biosphère et substituer à cet ancien paradigme un nouveau, c'est le mot "vivant" que nous avons choisi. "Tout est vivant, tout est lié" a été ainsi une des phrases les plus communément reprises du film. "Tout est nature, tout est lié" pourrait sembler très proche. Cette phrase n'a pourtant pas de résonance. Elle n'a en effet pas de sens: le mot "nature," avec son poids cognitif dans

[3] *HOME*, réalisé par Yann Arthus-Bertrand, produit par Elzevir Films et Europacorp, diffusé sur Internet gratuitement et dans les cinéma, les télévisions et DVD, la première fois le 5 juin 2009.

notre inconscient collectif nous renvoie à l'idée de la séparation et non à celle du lien.

Dans le même ordre d'idée, nous avons choisi dès les premiers mots de nous adresser au spectateur comme *Homo sapiens*. La seconde phrase du film, "tu es comme moi un Homo sapiens, l'homme qui pense," déplace le spectateur dans son identité. En adoptant cette rhétorique, nous tentons de le placer à un autre niveau qu'auquel il se place d'ordinaire: de l'interpeller comme espèce parmi les autres espèces, et aussi comme espèce douée de responsabilité parmi les autre espèces. Dans le changement de paradigme auquel nous souhaitions contribuer, l'individu n'est plus seulement un homme social parmi d'autres hommes de la même ou d'autres sociétés, il est aussi une espèce intégrée dans un Vivant dont il est à la fois issu, dépendant et maintenant responsable. Dans notre vision du monde, il faut sauter d'un cran sur notre vision de nous-mêmes.

Pour éviter la catastrophe, le philosophe Jean-Pierre Dupuy dans son ouvrage *Pour un catastrophisme éclairé* (2002) propose d'adopter la rhétorique de la prophétie: nous projeter dans la catastrophe annoncée pour pouvoir l'éviter. En dressant l'état des lieux de la planète et en donnant à voir les conséquences écologiques et sociales de notre vision du monde axée sur la prédation de la ressource, c'est cette méthode que nous avons plus ou moins adoptée dans *HOME*. Aujourd'hui projeté dans le monde entier et vu par un demi milliard de personnes sur la planète, *HOME* semble donner raison à J.P Dupuy: deux ans après sa première diffusion, le film ne cesse d'être relayé et d'être vu par un nombre croissant de personnes. Le langage adopté fondé en premier lieu sur la beauté des images d'une Terre-maison à préserver et combiné dans un second temps avec des informations alarmantes sur son état, provoque une prise de conscience.

Néanmoins, dans mes contacts avec le public et dans le suivi des répercussions du film, j'ai pu constater une limite majeure à cette méthode: montrant l'ampleur et à la multiplicité de la menace, elle peut entraîner aussi le déni et la paralysie chez nombre de personnes et finalement ne convaincre que les convaincus. Il me semble en fait que si la plupart des personnes ont profondément compris la situation, beaucoup d'entre elles refusent de l'accepter et restent dans le déni. Ce qui leur manque pour accepter la réalité est la projection dans un monde qui pourrait se structurer autrement. La rhétorique prophétique de la catastrophe semble ainsi limitée pour engendrer l'action et participer réellement à un changement de la vision de ce que pourrait être notre monde. Il n'y a peut-être pas de changement de paradigme possible sans vision de ce que peut être ce nouveau paradigme, ou dit autrement, sans vision de ce que peut être cette nouvelle vision du monde: c'est à dire, finalement sans *fiction*.

Dans son article "Du bon usage des catastrophes, pour *Le Monde de l'éducation* paru en 2006, J.P.Dupuy prône l'apprentissage de certaines compétences pour appréhender et éviter la catastrophe, dans les disciplines scientifiques comme humaines "du côté des sciences je distinguerai [...] l'apprentissage de la complexité, du côté des humanités, la pratique assidue de la capacité narrative" (40-43). Pour éviter la catastrophe,

> le catastrophisme éclairé est une ruse qui consiste à faire comme si nous étions la victime d'un destin tout en gardant à l'esprit que nous sommes la cause unique de ce qui nous arrive. Or cette ruse, ce double-jeu, la pratique de la fiction littéraire nous amène à y avoir recours le plus naturellement du monde. [...] La compétence requise est bien celle de faire et de comprendre des récits, et rien mieux qu'une formation littéraire classique ne peut permettre de l'acquérir. (43)

Cette capacité de narrer la fiction est tout autant importante pour la projection dans de nouveaux possibles, et pour nous permettre de les réfléchir et de les construire ensemble.

Les scientifiques ont besoin des narrateurs pour diffuser largement leurs connaissances dont l'appropriation est désormais devenue indispensable à tout citoyen qui souhaite comprendre le monde, saisir son évolution et œuvrer en adéquation avec ses changements. Les techniciens du changement en ont autant besoin pour diffuser les solutions complexes qu'ils mettent en œuvre tant au plan organisationnel que technologique afin que chacun puisse se projeter dans des règles du jeu qui deviennent différentes. La complexité nécessite d'inventer sa forme d'expression pour être appréhendable, elle nécessite d'être digestible, pour être transformable. C'est la *narration* d'un nouvel avenir possible qui nous permettra de le faire exister avant qu'il n'existe: en ouvrant un espace commun, mythique avant d'être réel, elle nous permettra d'y inscrire nos destins, nos vies et nos métiers, et ainsi peut-être, de le faire exister dans la réalité.

Si au premier regard, cette époque apparaît angoissante par les enjeux qu'elle demande de relever, au second, elle se révèle être tout autant une opportunité étonnante pour rassembler ce qui a été disjoint: l'homme et le vivant, les connaissances de la matière et celles des humanités.

Ouvages cités

Beauchamp, André. *Introduction à l'éthique de l'environnement*. Montréal: Éditions Paulines, 1993.

Dupuy, Jean-Pierre. "Du bon usage des catastrophes." *Le Monde de l'éducation*, 349, juillet-août, 2006: 40-43.

______. *Pour un catastrophisme éclairé. Quand l'impossible est certain*. Paris: Éditions du Seuil, 2002.

Guérin, Emmanuel et Michel Colombier. *Developed Countries 2020 Pledges Fall Short of IPCC Target. What can we do? Synthèses* 5. IDDRI. (09 décembre 2009). <http://www.iddri.org/Publications/Collections/Syntheses/sy_0905_Guerin_Colombier_Dev_Pledges.pdf>

HOME, film réalisé par Yann Arthus-Bertrand, produit par Elzevir Films et Europacorp, 2009.

Jackson, Tim. *Prospérité sans croissance. La transition vers une économie durable*. Trad. Patrick Viveret. De Boeck/Etopia, 2010.

Lorius, Claude, et Laurent Carpentier. *Voyage dans l'anthropocène: cette nouvelle ère dont nous sommes les héros*, Arles: Actes Sud, 2011.

Morin, Edgar. *La voie: pour l'avenir de l'humanité*. Paris: Fayard, 2011.

Pachauri, Rajendra K. et Andy Reisinger, dir. *Bilan 2007 des changements climatiques: Rapport de synthèse*. 4e rapport du Groupe d'experts Intergouvernemental sur l'évolution du climat. Genève: GIEC, 2007. <http://www.ipcc.ch/pdf/ assessment-report/ar4/syr/ar4_syr_fr.pdf>

CPSIA information can be obtained at www.ICGtesting.com
Printed in the USA
LVOW05s1131100813
347201LV00001B/57/P

9 789042 036130